Computational Fluid Dynamics Applications in Bio and Biomedical Processes

Satya Eswari Jujjavarapu · Tukendra Kumar · Sharda Gupta

Computational Fluid Dynamics Applications in Bio and Biomedical Processes

Biotechnology Applications

Springer

Satya Eswari Jujjavarapu
Department of Biotechnology
National Institute of Technology
Raipur (NIT)
Raipur, Chattisgarh, India

Tukendra Kumar
Department of Biotechnology
National Institute of Technology
Raipur (NIT)
Raipur, Chattisgarh, India

Sharda Gupta
Department of Biomedical Engineering
National Institute of Technology
Raipur (NIT)
Raipur, Chattisgarh, India

ISBN 978-981-99-7128-2 ISBN 978-981-99-7129-9 (eBook)
https://doi.org/10.1007/978-981-99-7129-9

This Springer imprint is published by the registered company Springer Nature Singapore Pte Ltd.
The registered company address is: 152 Beach Road, #21-01/04 Gateway East, Singapore 189721, Singapore

Paper in this product is recyclable.

About This Book

Computational fluid dynamics (CFD) tools play a pivotal role in simulating and analysing diverse biochemical and physical behaviours, optimizing designs, and enhancing efficiency in engineering and scientific applications. The book comprehensively covers all aspects of CFD, showcasing the latest applications in biomedical, biotechnological, and bioelectrochemical engineering. With its wide-ranging applications, this book meets the demand for novel designs and bioprocess optimization, leading to cost reduction and decreased experimental burden. In response to the growing demand for value-added products, this book helps industries must actively seek novel designs to improve production efficiency. Furthermore, the book highlights the game-changing application of CFD in biomedical research, enabling a deeper understanding of complex biological processes, enhancing medical device design, and optimizing drug delivery strategies. As CFD continues to advance, its impact on biomedicine is bound to grow, promising transformative breakthroughs in healthcare and medical technologies. In essence, this book serves as a guiding beacon, empowering researchers to explore new frontiers and foster innovative, sustainable biotechnological advancements that positively impact human health and the well-being of our planet. By harnessing the power of CFD, researchers and industries alike can revolutionize their approach, leading to a more efficient and impactful future.

Contents

About the Authors

Dr. Satya Eswari Jujjavarapu is currently an assistant professor at the Biotechnology Department of the National Institute of Technology (NIT), Raipur, India. She did her M.Tech in Biotechnology from the Indian Institute of Technology (IIT) Kharagpur and her Ph.D. from IIT, Hyderabad. During her research career, she worked as a scientist (woman scientist—Department of Science and Technology (DST)) at the Indian Institute of Chemical Technology (IICT) Hyderabad. She has published more than 75 SCI/Scopus research papers, 2 patents, 9 books, few book chapters, and 45 in international conference proceedings. Her research contributions have received wide global citations. She completed three sponsored projects. She has been teaching experience of more than ten years and three years of research experience. She received awards as a guest editor for the journal *Indian Journal of Biochemistry and Biophysics* (SCI), twice *Journal of Chemical Technology and Biotechnology*, received with IEI Young Engineer Award-2019, Early Career Research Award, Outstanding Woman by Venus international-2019, DK Best Faculty award-2019. INAE Nominated twice. She has rigorously pursued her research in the areas of environmental bioremediation, wastewater treatment, bioprocess, product development, and bioinformatics. She gained pioneering expertise in the application of mathematical and engineering tools to biotechnological processes.

Mr. Tukendra Kumar is currently a senior research fellow in the Department of Biotechnology at the National Institute of Technology (NIT), Raipur, India. Mr. Kumar has been awarded the prestigious DBT JRF fellowship and secured a coveted Ph.D. program fellowship from the esteemed Department of Biotechnology, Ministry of Science and Technology, Government of India. He has remarkable achievements in various national-level examinations. He achieved outstanding scores in CSIR NET 2019 life science, GATE Biotechnology, and Life Science. His academic journey began at Guru Ghasidas Vishwavidyalaya (GGV), Bilaspur, India, where he earned his post-graduation in biotechnology. Prior to that, he laid the foundation of his academic excellence at Pt. Ravishankar Shukla University (PRSU), Raipur, Chhattisgarh, India, where he pursued his undergraduation in Biotechnology. His work has been recognized in multiple international peer-reviewed journals, with five publications to his credit. Additionally, he has been actively involved in sharing his insights at renowned international conferences through several conference proceedings. His dedication to innovation and discovery is further reflected in the two Indian patents currently under review. His major research areas and interests are bioelectrochemistry, biosensors, environmental bioremediation, computational fluid dynamics, and bioprocesses.

Dr. Sharda Gupta is currently a contract faculty at the Biomedical Department of the National Institute of Technology (NIT), Raipur, India. She is the recipient of the Chhattisgarh Young Scientist Award 2023. She has received doctorate degree (2022) from the National Institute of Technology, Raipur. She is a GATE fellow (2012) and was a university topper in her master degree (2014) from the School of Instrumentation, DAVV Indore. In her master's project, she worked on temperature monitoring and data logging in CRYOGENICS Using LabVIEW at Cryoengineering and Cryomodule Development Division (CCDS), RRCAT INDORE, Department of Atomic Energy, Government of India. She has authored more than ten articles in peer-reviewed journals, seven book chapters, and two patents are under review. She is actively engaged in multidisciplinary research involving frontier areas of skin tissue engineering, 3D skin disease models, microfluidic chips, use of silk-based biomaterials, 3D printing and implementation of electronics in the field of biology in the form of biosensors.

Chapter 1
Computational Fluid Dynamics: Fundamentals and Applications in the Design and Optimization of Various Bioreactors

1 Introduction

Traditional design methods that rely on empirical correlations or physical experiments frequently limit and lengthen design iterations. Computational fluid dynamics (CFD) emerges as a powerful instrument for simulating and analysing biochemical properties within bioreactors in order to overcome these obstacles. This chapter discussed the fundamental concepts and methodologies of CFD, such as numerical methods, and turbulence modelling techniques. The applications of CFD in activated sludge systems, anaerobic digesters, membrane bioreactors, and other forms of bioreactors are also discussed. This analysis found that the optimization of design parameters including reactor geometry, aeration systems, impeller configurations, and operational conditions improved the efficiency of bioreactors. In addition, CFD facilitates the evaluation of various design scenarios, the troubleshooting of extant bioreactors, and the prediction of system performance under varying operating conditions. Additionally, this chapter emphasizes current and future challenges and limitations in applying CFD to bioreactor design, such as the requirement for precise boundary conditions, model validation, and computational resources. CFD enables enhanced performance, efficacy, and cost-effectiveness of bioreactor systems by providing a comprehensive understanding of flow phenomena. This review paper is beneficial for researchers, engineers, and practitioners involved in bioreactor design and optimization in a variety of industries.

2 Background

Bioreactors are essential in a variety of industries, including wastewater treatment, biogas production, and the creation of bio-based products. These constructed systems provide a setting for biological processes that involve microorganisms interacting

S. E. Jujjavarapu et al., *Computational Fluid Dynamics Applications in Bio and Biomedical Processes*, https://doi.org/10.1007/978-981-99-7129-9_1

with organic matter or other substrates. When compared to typical batch operations, bioreactors have advantages such as improved process control, scalability, and higher product yields. However, due to complex fluid dynamics, mass transfer phenomena, and the delicate interplay between biological and physical processes, building and optimizing bioreactors present significant obstacles (Mandenius, 2016; Song et al., 2018).

Understanding and regulating the fluid flow patterns within the system is one of the most difficult tasks in bioreactor design. It is critical to have appropriate mixing and effective mass transfer in order to ensure homogeneous distribution of nutrients, gases, and microbes. Inadequate mixing can cause localized substrate gradients, resulting in decreased microbial activity and unequal treatment efficiency (Nadal-Rey et al., 2021). On the other hand, excessive mixing can create shear stress on microorganisms, reducing their viability and limiting process performance. As a result, optimizing flow patterns and mixing regimes are critical for increasing the efficiency of bioreactor systems (Gómez-Ríos et al., 2019). In addition to fluid dynamics, effective mass transfer is crucial for bioreactor performance. The supply of oxygen, nutrients, and other important substances to microorganisms, as well as the elimination of metabolic waste, has a substantial impact on the biological processes' growth, activity, and productivity. Inefficient mass transfer can result in oxygen restrictions, substrate gradients, and the accumulation of inhibitory byproducts, all of which can have a negative impact on bioreactor efficiency (Pilarek et al., 2018). Furthermore, the intricacy of biological processes complicates bioreactor design even further. Nonlinearities and unpredictability are introduced into the design and optimization process by interactions between microbes, substrate utilization kinetics, and metabolic pathways. It is a huge problem to develop appropriate mathematical models to capture these intricacies and integrate them into the bioreactor design (Xu et al., 2018).

To address these issues, a thorough understanding of fluid dynamics, mass transport events, and the underlying biological processes within bioreactors is required. CFD appears as a useful technique in this situation. CFD simulations aid in the optimization of bioreactor design and operational parameters by allowing the examination of fluid flow patterns, mixing efficiency, and mass transfer rates (Kumar & Jujjavarapu, 2023a). CFD has significant economic and time implications for better wastewater system design and biomass utilization (Climent et al., 2018). Due to the aeration system, biological reactors are the most energy-intensive facilities of conventional wastewater treatment plants (WWTP). Numerous biological reactors use intermittent aeration; optimizing the aeration process is required to ensure high efficiency, meeting untreated sewage requirements while using the least amount of power (Gu et al., 2023). CFD has significant financial and risk drivers for improved wastewater system design for biomass utilization and waste treatment.

The goal of this chapter was to provide a complete overview of the use of CFD in the design and optimization of various bioreactors used in wastewater treatment, biogas production, and the manufacture of bio-based products. The review used a comprehensive literature search to identify and analyse relevant research papers, journals, and conference proceedings. The selected papers were critically reviewed in order to acquire information on the use of CFD in bioreactor design and optimization, including case studies, developments in CFD techniques, and field problems. It sheds light on the significance of CFD in understanding fluid dynamics, mass transport phenomena, and process optimization in bioreactors. The analysis produced a thorough overview of CFD applications, highlighting important accomplishments, problems, and future prospects in the discipline. It was a helpful resource for bioreactor design and optimization researchers, engineers, and practitioners from a variety of sectors.

3 Computational Fluid Dynamics

CFD, including its diverse applications, has been commonly used in manufacturing and scientific studies since its inception in the mid-twentieth century. CFD is an intriguing discipline within the realm of fluid mechanics, focusing on the intricate numerical simulation of fluid flow and the captivating phenomena of heat transfer. This captivating field harnesses the power of computational methods to unravel the mysteries of fluid dynamics and delve into the intricacies of heat transfer. The fascinating realm of CFD revolves around harnessing the power of computer algorithms and numerical methods to tackle the intricate governing equations that elucidate the intricate behaviour of fluids (D'Bastiani et al., 2023; Lee et al., 2013; Xu et al., 2010). As a result, CFD is used for the simulation and modelling of a variety of processes such as fluid flow, velocity, and pressure. In the CFD study, computer systems are required to conduct the computations mandated to simulate the fluid surface interaction characterized by boundary conditions (Zawawi et al., 2018). High-performance supercomputers could be used to obtain faster and better solutions. The Navier–Stokes equations are the basic model of almost all CFD problems and characteristics of nearly whatever single-phase fluid flow. CFD has contributed in so many industry sectors in the design and analysis environments because of its potential to predict the progress of modern designs or structures before they are constructed or implemented (Alobaid et al., 2022; Shen et al., 2020).

Numerical methods, a cornerstone of scientific and technological advancements, encompass a range of mathematical techniques employed to tackle intricate equations or problems that elude analytical solutions. In the realm of CFD, the utilization of numerical techniques becomes imperative to discretize the continuous governing equations governing fluid flow. This process involves transforming these equations into a collection of algebraic equations, which can subsequently be solved through numerical methods on a computer.

3.1 Numerical Method in CFD

Various numerical solutions have emerged over the last few years. There are various numerical discretization strategies in algebraic equations and computational mathematics. There are five basic numerical methods in CFD numerical solution, i.e. finite volume method (FVM), finite difference method (FDM), finite element method (FEM), lattice Boltzmann method (LBM), and smoothed particle hydrodynamics (SPH) (Afshari et al., 2018; Hassanzadeh et al., 2018). However, the finite volume method is the most widely used solver method for CFD. The important numerical approaches that are routinely employed in CFD are discussed as follows.

3.1.1 Finite Volume Method

The FVM, also known as the control volume method, is a common approach for discretization. The finite volume method solves the governing partial differential equation over finite control volumes. The fundamental concept of the FVM is to divide the entire domain into small control volumes. These volumes swap fluxes to their neighbours across the linking face. Because the fluxes which enter one cell leave another, this method is solely conservative, making it very appealing for flow problems. It is primarily derived from conservation-type calculations, and it makes an integral in its control volume before solving the conservation equation in integral form (Jasak & Uroić, 2020; Muhammad, 2021).

Jasak and Tuković (2010) provided a comprehensive description of the discretization process employed in the FVM, highlighting its key properties. The conservation equations are discretized using control volumes in the integral form, thereby guaranteeing the preservation of quantities such as mass and momentum at discrete levels. The FVM provides the advantage of being able to adapt to different control volume shapes, thus allowing for the consideration of a wide range of geometries. Furthermore, the boundary value problem is approached in a segregated fashion, wherein distinct aspects are dealt with individually. The utilization of a sequential approach facilitates the process of problem-solving by reducing complexity. Jasak's research offers a comprehensive elucidation of the discretization process employed in the FVM, emphasizing the significance of conservation principles, the adaptability of control volume geometries, and the segregated approach as a means to enhance computational efficiency.

Let us consider a computational point P situated at the centroid of a control volume. In the given configuration,

$$\int_{Vp} (X - Xp)dV = 0. \tag{1}$$

In this context, the symbol *Vp* is used to represent the volume of the control volume; *Xp* is employed to indicate the centroid of the control volume while *dV* represents a small volume within the control volume (Muhammad, 2021).

3.1.2 Finite Difference Method

The finite difference method (FDM) is the oldest method of computer simulation analysis, and it works well with orderly and structured geometries. The FDM discretizes the Navier–Stokes equations on a number of points which represent the fluid's physical parameters, such as velocity and pressure (Udoewa & Kumar, 2012). These points are generally dispersed on a continuous basis because it makes application much simpler. A Taylor expansion is used to calculate the derivative at grid nodes. However, with its reliance on grid points, the FDM is difficult to apply to problems with complicated boundaries. But still it could be used, particularly in applications where only a regular grid is used, such as weather simulations (Bhanduvula, 2012; Ye et al., 2022).

Using finite difference formulas, the FDM approximates the derivatives in the governing equations. The general discretization equations for a 1D problem are as follows:

- Discretization of spatial domain:

$$\Delta x = \frac{(x_{\max} - x_{\min})}{N - 1}, \tag{2}$$

  where Δx is grid spacing and N is the number of grid points.
- Discretization of governing equations:

  (a) Conservation equation:

$$\frac{\partial \varphi}{\partial t} + \frac{\partial F}{\partial x} = 0. \tag{3}$$

  (b) Diffusion equation:

$$\frac{\partial(\rho\varphi)}{\partial t} = \partial\frac{(D\partial\varphi/\partial x)}{\partial x} \tag{4}$$

- Boundary conditions:

  (a) Dirichlet boundary condition:

$$\varphi(x_{\min}) = \alpha, \varphi(x_{\max}) = \beta \tag{5}$$

  (b) Neumann boundary condition:

$$\frac{\partial \varphi}{\partial x} = y \text{ at } x = x_{\max/\min}. \tag{6}$$

- Time stepping:
  Forward Euler method:

$$\varphi i(n+1) = \varphi i(n) + \Delta t * \left(\frac{\partial \varphi}{\partial t}\right) * i(n) \tag{7}$$

3.1.3 Finite Element Method

The finite element method (FEM) is commonly used in the analysis of solid structures, although it is appropriate for fluids. In general, this method depends on the variation concept and the weighted margin domain. Because the domain is divided into several elements, this integration occurs over every element in the mesh. Piecewise polynomial features estimate the solution in each of these elements. These are defined by a set of basic functions for each nodal point and could then be used to approximate derivatives (Abali, 2019; Dick, 2009a, 2009b, 2009c). However, the FEM necessitates more memory than the FVM but this method is more stable.

3.1.4 Lattice Boltzmann Method

The lattice Boltzmann method (LBM) is a linear and effective algorithm for modelling single-phase and multiphase flows, as well as integrating significant extra difficulties and challenges (Arumuga Perumal & Dass, 2015). The LBM is a promising new choice to conventional CFD methods. The Boltzmann equation is utilized in this approach to estimate macroscopic fluid circumstances as mesoscopic shock developments of fictitious particles. The LBM is particularly useful for simulating complex boundary conditions and multiphase interfaces. Since this strategy is not based on Navier–Stokes equations, several people still assume it is not a CFD method (Glessmer & Janßen, 2017; Mehl et al., 2011).

3.1.5 Smoothed Particle Hydrodynamics

Smoothed particle hydrodynamics (SPH) is a quick and reliable numerical technique for solving a wide range of difficult problems in CFD. This is a strategy that is suitable for deformation problems (Shadloo et al., 2016). To simulate fluid flow, this method also discretizes using Navier–Stokes equations. In contrast to the other methods, the SPH method is a Lagrangian mesh-free discretization method. The properties of the system are described by a set of particles in the SPH approach, each of which has individual material characteristics and interacts with one another within a particular range described as a support domain by a weight function or smoothing function.

Single flow and multiphase flows are also simple to handle by SPH because the moving interface between the phases is automatically tracked. Moving boundaries and free surface flows are also efficiently handled (Glessmer & Janßen, 2017; Korzani et al., 2016). As a result, the SPH strategy allows the flow simulation of previously impossible-to-simulate use cases.

The various numerical methods offer distinct trade-offs in terms of precision, stability, and computational efficacy. The selection of a particular methodology is contingent upon several factors, including the inherent characteristics of the problem at hand, the spatial configuration of the domain, and the computational capabilities that are at one's disposal. Furthermore, the integration of turbulence modelling, grid generation, and time-stepping schemes is employed in conjunction with these numerical methods to effectively address specific CFD challenges with precision.

3.2 CFD Software and Overview of Modelling

CFD software allows researchers to evaluate fluid flow effects on surfaces directly on your computer. CFD can easily test, simulate, and solve your problems and equations, and you can ensure that your design or product meets all safety and structural integrity standards (Zawawi et al., 2018). Some of the most commonly used CFD software are listed below:

- COMSOL Multiphysics
- Autodesk CFD
- IVRESS
- Altair HyperWorks
- Power FLOW
- FLOW-3D
- Ingrid Cloud
- ANSYS Fluids
- AUSM
- Avizo (software)
- FLACS
- OpenFOAM
- TELEMAC
- ADINA.

CFD analysis is often divided into four stages (Fig. 1) (Bhaskaran & Collins, 2020; Minin, 2011; Mrope et al., 2021):

(i) Problem identification: In this step, the problem is explicitly defined, and the objectives of the CFD study are established. Geometry, boundary conditions, beginning conditions, and any other important characteristics are all specified. At this step, it is critical to understand the physical events and select appropriate models and equations to reflect the flow behaviour.

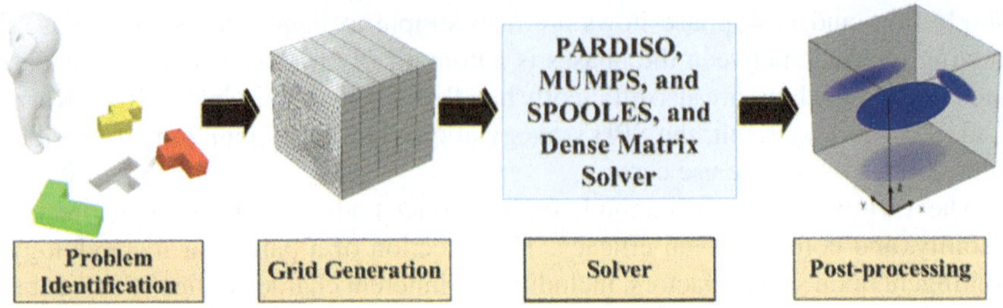

Fig. 1 Overview of CFD modelling

(ii) Grid generation: A computational grid or mesh is generated in this step to discretize the issue domain. The grid might be structured (e.g. a Cartesian grid) or unstructured (e.g. triangular or tetrahedral pieces). It is critical to generate a suitable grid that accurately reflects the geometry and flow properties in order to produce reliable findings. The density and quality of the grid can have a considerable impact on the accuracy and computing efficiency of the study.

(iii) Solver: The governing equations (such as the Navier–Stokes equations for fluid flow) are discretized using numerical methods once the grid has been constructed. This entails approximating the derivatives and solving the algebraic equations that result on the grid. Finite difference, finite volume, and finite element approaches are common numerical methods used in CFD. To reach a converged solution, various solution approaches such as iterative solvers and time-stepping methods are used.

(iv) Post-processing and analysis: These are carried out after the numerical solution has been obtained. This stage entails obtaining useful data from the results, such as velocity profiles, pressure distributions, and heat transfer rates. Visualization techniques such as contour plots, streamline plots, and animations are frequently utilized to efficiently comprehend and present the results. To assess the correctness and reliability of the CFD analysis, sensitivity analyses, validation, and verification methods may be performed.

It is crucial to remember that these phases are interrelated and iterative. On the basis of the outcomes and new information learned throughout the process, the analysis may need to be refined or have some portions revised.

4 Computational Fluid Dynamics for Design and Optimization of Reactors

CFD is commonly utilized in wastewater treatment reactors to solve complex problems. CFD is increasingly being used in the development of design features and process optimization through computer analysis, resulting in lower design, and operating costs and increased efficiency of the reactors. CFD is a useful tool for reactor

Table 1 Commonly used CFD models and governing equations for bioreactors

CFD model	Formula/equations	References
Eulerian–Eulerian model	Navier–Stokes equations for each phase	Aly et al. (2010), Suh et al. (2018)
Eulerian–Lagrangian model	Navier–Stokes equations for the continuous phase and particle tracking equations for the dispersed phase	Balakin et al. (2010), Le Lee and Lim (2017)
Two-fluid model	Conservation equations for each fluid phase	Jamshidian et al. (2023), Werner et al. (2014)
RANS model	Reynolds-averaged Navier–Stokes equations	Bach et al. (2017), Wu (2013)
LES model	Filtered Navier–Stokes equations with subgrid-scale models	Cantarero Rivera and Chen (2022), Li et al. (2018a, 2018b)
Population balance model	Population balance equation for particle size distribution	Farzan et al. (2017), Morchain et al. (2013)
Biofilm models	Biofilm growth and detachment equations	Yao et al. (2022)
Porous media models	Darcy's law and mass transfer equations for porous media	Wang et al. (2010), Weyand et al. (2015)

design and optimization. It allows for the analysis of fluid flow, heat transfer, and reaction kinetics within reactors, resulting in better performance. For the design and process optimization of the bioreactor, various types of CFD models can be utilized (Table 1). CFD plays an important role in enhancing reactor design and optimization in a variety of industries, including chemical engineering and energy generation, by offering deep insights into flow behaviour and reaction dynamics (Singh & Hutmacher, 2009). The following are some of the applications of CFD in the most common types of reactors.

4.1 Activated Sludge Bioreactors

The activated sludge process is a multichamber reactor unit which employs extremely concentrated bacteria to degrade organic material and eliminate contaminants from wastewater, resulting in high-quality effluent (Wang et al., 2018). One of the significant issues in the sewage treatment sector is the effective configuration and higher reliability of treatment operations that ensure increased treatment efficiency levels to meet sludge quality requirements while keeping the cost of investment as minimal as possible. Although design concepts and processing methods of activated sludge plants are critical to achieving these objectives, they are still based on basic experimental method (Karpinska & Bridgeman, 2016a).

The biological processes, such as the biodegradation of pollutants by microorganisms in an activated sludge reactor (ASR), depend on oxygen concentration to maintain an aerobic condition. A most essential factor in ASR designs is the improvement of oxygen mass transfer efficiency, which is accomplished by aeration through diffusers in the channel or closed-loop bubbly flow in ASR. Higher biodegradation efficiency of ASR can be achieved by increasing the surface area of the interface between the oxygen (air) bubbles phase and the liquid phase. CFD can be used by several researchers as an advanced construction method for the advancement of novel aeration devices as well as the optimization of their spatial framework in ASR. The sk $-$ ε turbulence model is the most commonly used model for studies of single-phase simulation in three-dimensional steady state. Bhuyar et al. (2009) used a single-flow analysis strategy to change the configuration and operating condition of a curved blade aerator designed for use in oxidation ditches. When compared to traditional mechanical devices, the authors' findings revealed that optimized blade design and aerator improved aeration efficiencies. The computational result was validated through the experiment, which found good agreement with the CFD result.

Even with the greater computational cost, the transitory gas–liquid computer simulation method which includes unsteady Reynolds-averaged Navier–Stokes simulation with an Euler–Euler multiphase model and sk $-$ ε turbulence model is still used in the advancement of novel aeration strategies because it provides a more precise and factual assessment of multiphase reactor systems. Xu et al. (2010) used this modelling method to simulate an oxidation ditch with cylindrical airlift aerators. CFD gas–liquid simulation method for liquid flow in an airlift oxidation ditch has been used in this case to validate the feasibility of the preliminary design and assess its suitability for sewage treatment. The simulation studies emphasized the suggested aeration device suitable for deep tank ditch designs. A similar model was used by Norouzi-Firouz et al. (2022) to simulate coupled hydrodynamic-biokinetic operations in an activated sludge oxidation ditch method. The authors investigated the influence of fluid flow variables on the efficiency of a novel aerator and validated their findings against experimental results. They revealed that the quantity of aerators has an overt impact on baffle efficiency and the fluid phase velocity distribution.

Recently, Matko et al. (2021) used CFD two-phase simulation modelling for an aerated wastewater treatment oxidation ditch, accounting for gas–liquid flow, oxygen mass transfer, and dissolved oxygen. These authors investigate the effect of the bubble size distribution and the biochemical oxygen demand distribution on the total dissolved oxygen distribution in the aerator system (Fig. 2). The bubble size distribution predicts an average bubble size, i.e. 2 mm and the result found this bubble size best with DO measurements. This investigation also explores the biochemical oxygen demand distribution and BSD as essential factors for the consistency of the total dissolved oxygen distribution that has largely been ignored in scientific research. Similarly, Xu et al. (2021) developed a hydrodynamic-biokinetic model that combines CFD and ASM2 to optimize operational methods in oxidation ditches. It solves OD limits in wastewater treatment and provides a thorough simulation of hydrodynamics and biokinetics. The model provides insights into flow behaviour, suspended solids distribution, and biological reactions, enabling energy expenditure

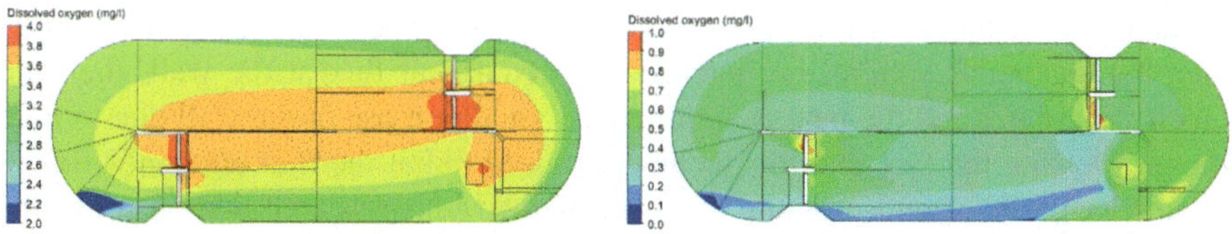

(a) (b)

Fig. 2 Dissolved oxygen distribution without (**a**) and with BOD distribution and BSD (**b**) (Matko et al., 2021)

and operating cost optimization. It is an important tool for OD management and can be used to increase the efficiency of wastewater treatment plants.

4.2 *Fixed-Bed Reactors*

Fixed bed, fixed film, or attached growth reactors are biological wastewater treatment processes that use an inert medium such as rock, plastic, wood, or other natural or synthetic solid material to support the attached growth of active biomass responsible for degradation on the surface and within the porous structure reactors. Wastewater treatment using a fixed-fed reactor is the second most popular method for removing contaminants from water. However, the optimum condition required for the increase of the efficiency of film reactors for the removal of contaminants from wastewater (Hickman et al., 2016). Logtenberg and Dixon (1998) used ANSYS/FLOTRAN CFD software to solve and optimize the 3D Navier–Stokes and energy equations for fluid and heat transfer in a fixed bed reactor. They used a 2D pseudo-homogeneous model with a structure of eight spheres and boundary conditions at the wall and inlet to assess the temperature and velocity at different spots inside the reactor. The results showed that using CFD in fixed bed tubes to obtain values for heat transfer coefficient (N_{uw}) and thermal conductivity ratio (kr/kf) was effective and approximately the same as an experimental value. It can be eventually concluded that CFD is a viable tool for assessing heat transfer behaviour in a fixed bed reactor. Dixon and Partopour (2020) summarized the progress of fixed bed reactors by particle-resolved CFD simulations, which began in the mid-1990s. Many crucial aspects have been completely covered, such as packing generation, meshing, and applications of fluid dynamic challenges like heat transfer, mass transfer, and chemical reactions. Because computational equipment was restricted at the time, the research was constrained either to periodic segments of a frequently organized packing or a small number of particles constituting a random fixed bed.

In the last ten years, computer hardware has become faster and much more affordable. This has resulted in an increased interest in employing CFD to optimize the reactors in the sector of industrial and academic areas (Jurtz et al., 2019). Yang et al. (2016) investigated the impacts of fixed bed geometry parameters such as particle size and wall distance on fluid flow and heat profiles. The results demonstrate that when homogeneous spheres are packed to a fixed heat flux at the wall, the fluid velocity is greater and the heat is significantly smaller close to the wall due to high voltage. Despite the fact that this technique raised radial thermal efficiency due to the smaller particles and pressure drop accelerated, this research demonstrates the use of addressed particle CFD simulations for design and optimization. Zhang et al. (2018) performed a study for a homogeneous random manner packed particle beds reactor, which revealed similar fluid motion and thermal gradients close to the wall-to-sphere packing. In addition, the researchers presented a better 2D uniform model based on the outcome of their 3D investigated particle model. Dong et al. (2017) used new profile reactor monitoring results to verify their resolved particle CFD analysis. Although the author's simulation results did not consider the sampling capillary utilized in the laboratory experiments assessments.

Numerous recent research has also utilized CFD analysis for the fixed bed to further enhance the large industrial reactions like methane steam, and ethylene oxidation. By utilizing resolved particle simulation studies, G. M. and Buwa (2018) investigated fixed beds under methane steam reforming situations. These experiments used a small amount of particles and a relatively short, fixed bed length. According to the findings of the research, shaped particles like daisies and trilobed decrease pressure drop and boost distribution in the methane steam reforming fixed bed. Singhal et al. (2017) conducted a similar study utilizing resolved particle simulations to fix beds under methane steam reforming conditions. However, the analysis was used by these authors to enhance their 1D packed bed model, which was then applied to packed bed chemical looping reforming. Partopour and Dixon (2016) used resolved particle CFD to investigate ethylene partial oxidation to ethylene oxide. The research illustrates the fault of reduced microkinetics models inside the case of large temperature and species gradients in 3D CFD simulations of fixed bed reactors. The study concentrated on the influence of temperature gradients on specificity and surface oxygen characteristics along this lengthy bed of 800 spherical particles. The analysis indicates that surface oxygen is closely linked with the temperature gradient within the fluid and solid particles, and therefore, it regulates reaction specificity.

One study intends to improve the efficiency of fixed bed reactors for removing contaminants from wastewater by designing and optimizing the reactor using CFD simulation. Sandhibigraha et al. (2019) optimized an immobilized catalysed fixed bed reactor for the removal of 4-chlorophenol from wastewater using the CFD ANSYS Fluent software. They optimized various operating variables such as 4-chlorophenol (4-CP) inlet concentration, fluid velocity, bed length, and pore size distribution to estimate the optimal 4-CP degradation process in a fixed bed reactor. The highest biodegradation of 4-CP was anticipated by a CFD simulation in the presence of 500 mg/L inlet concentration of 4-CP, 4 mL/min fluid velocity, 18 cm bed height, and 0.375 porosity. According to our information, there have been few studies published on the estimation of optimum parameter characteristics for biodegradation and pollutant removal using CFD analysis.

4.3 Membrane Bioreactors

In advanced wastewater treatment systems, membrane bioreactors (MBRs) combine biological processes with membrane filtration. MBRs use specialized membranes to separate solid particles, microorganisms, and dispersed particulates from purified water, resulting in effluent of superior quality. MBRs provide numerous benefits, such as a compact design, efficient particulate separation, a smaller footprint, and high-quality treated water. The membrane barrier offers superior solid–liquid separation and contributes to achieving low turbidity and the removal of microparticles, pathogens, and contaminants. MBRs have the ability to remove nutrients and are effective in treating a wide variety of wastewater types, including municipal, industrial, and domestic effluents (Goswami et al., 2018; Yoon, 2015).

However, MBRs also present challenges, such as membrane contamination, which occurs when particles accumulate on the surface of the membrane, reducing permeability and necessitating regular cleansing or replacement. Membrane corrosion management and associated costs continue to be areas of active research (Xiao et al., 2019). In recent years, CFD has emerged as a valuable instrument for diagnosing problems, understanding design considerations, and investigating the hydrodynamics, flow regimes, and process configurations in MBR. Additionally, CFD is utilized to simulate fluid behaviour on the membrane surface and pores, to model mass transfer rates, and to predict contamination layer formation. These numerical modelling techniques provide valuable insights into the analysis of complex multiphase flow in municipal MBRs when correctly validated with experimental data (Jalilnejad et al., 2022).

For the hydrodynamics effect of fluids on membrane fouling, the most common model of CFD is single-phase laminar flow (Navier–Stokes equations) as shown below:

$$\frac{\Delta(\rho)}{\Delta t}+\left[\frac{\Delta(\rho u)}{\Delta x}+\frac{\Delta(\rho v)}{\Delta y}+\frac{\Delta(\rho w)}{\Delta z}\right]=0 \tag{8}$$

In a Cartesian coordinate system, ρ represents the fluid's density, t indicates time, and u, v, and w represent the velocity components along the x, y, and z angles, respectively.

The Eulerian–Eulerian model was used to simulate the two-phase flow, while the particle model was used to simulate the interphase contact area, which governs momentum transfer across the interface. In this modelling strategy, one phase (water) is treated as the continuous phase, while the other (air) is treated as the dispersed phase (Wu et al., 2018).

The activated sludge system exhibits non-Newtonian fluid behaviour, in which the viscosity is affected by a number of variables, including shear rate, temperature, mixed liquor-suspended solids concentration (MLSS), extracellular polymeric substances (EPS), and soluble microbial product (SMP) concentration (Liu et al., 2015; Ratkovich et al., 2013). Several models, such as the Eyring model, the cross model, and the Bingham plastic model, have been proposed to characterize the relationship between shear stress and shear strain rate in non-Newtonian fluids. Among these models, it was determined that the Ostwald-de Waele model best describes the behaviour of sediment, as expressed by the following equation (Liu et al., 2015):

$$\tau = 0.0431MLSS^{0.89}\gamma^{0.68}(3 \le MLSS \le 8). \tag{9}$$

$$\tau = 0.0412MLSS^{1.64}\gamma^{0.45}(8 \le MLSS \le 16). \tag{10}$$

Most of the researchers have employed CFD techniques in their investigations aimed at mitigating membrane fouling in MBRs. The location of membrane units within the membrane vessel is crucial to the design of an MBR at maximum capacity. It has a significant effect on the hydrodynamics within the tank, which in turn affects the corrosion rate of the membrane during operation. In a study, the hydrodynamics of the membrane tank in a full-scale MBR were simulated using an integrated CFD model (Newtonian and non-Newtonian fluid model). The fifth percentile of the liquid velocity (v0.05) in membrane units was used as the risk velocity to determine the contamination potential of the membrane. The optimal design for the unit location was proposed ($i_{Lu} = 0.6$, $i_{La} = 0.6$, $i_{Lb} = 0.6$, $i_{Lint} = 0.6$, and $i_{Lw} = 0.6$), with a risk velocity $v_{0.05}$ elevation of 146.9% (Wu et al., 2018). Turbulence promoters serve as an alternative strategy for enhancing the efficacy of membrane filtration and mitigating the occurrence of membrane fouling. A study investigated the use of a microchannel turbulence promoter (MCTP) in a submerged flat-sheet membrane bioreactor (SMBR) by CFD Euler multiphase model. In comparison to the horizontal orientation, the vertical orientation of the SMBR with MCTP demonstrated greater average fluid velocity, gas velocity, turbulent kinetic energy, wall shear stress, and gas holdup. With improvements ranging from 8.7 to 37.76%, the enhancement effect of the vertical orientation was preferable to that of the horizontal orientation. These results suggest that the hydrodynamic and filtration efficacy of the SMBR can be improved by incorporating MCTP, specifically in a vertical orientation (Xie et al., 2018).

In order to address the issue of uneven bubble distribution and subsequent membrane fouling in MBRs, Liu et al. (2019) developed a 3D CFD k-ε turbulent model and Ostwald-de Waele model. The aim was to predict the hydraulic performance of a novel aeration system designed to mitigate the problem. The results revealed that placing the aerator at a distance of approximately 20 cm from the membrane module, within the range of 15–25 cm, resulted in higher shear stress near the permeate collector (Fig. 3). This location exhibited higher flux and showed a maximum increase of 2.4 times in average shear stress. These findings suggest that optimizing the positioning of the aerator can lead to improved hydraulic performance and reduced membrane fouling in full-scale MBRs (Liu et al., 2019). The distribution of shear stress was found to become more homogeneous as the concentration of mixed liquor-suspended solids (MLSS) increased, resulting in a more uniform distribution of shear stress. The high shear rate and even bubble distribution was achieved through the development of a novel in situ aeration method based on the CFD Eulerian model and the k- ε turbulent model, and it was discovered that this method has superior membrane antifouling potential in the MBR (Shen et al., 2023).

Nonetheless, CFD modelling for MBRs still faces obstacles, such as capturing multiphase flow, developing a valid model for predicting membrane fouling, managing computational resources, parameterizing models, and scaling up. CFD in MBRs requires ongoing research to enhance its accuracy and applicability.

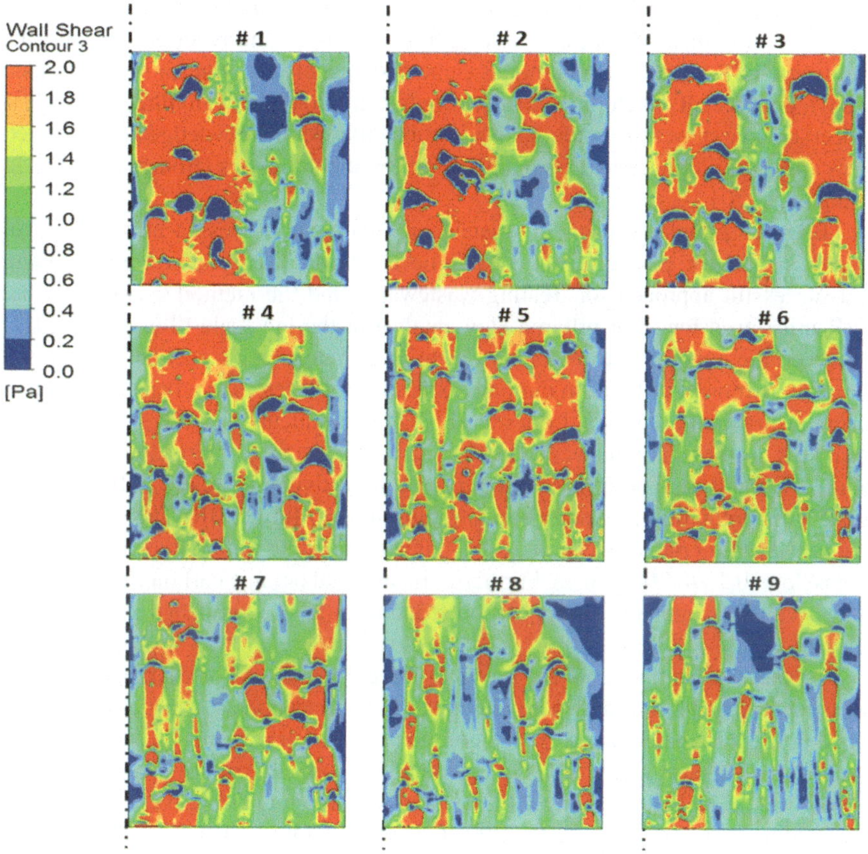

Fig. 3 Shear stress distribution on the surfaces of various membrane sheets for the standard scenario at 2.5 s (Liu et al., 2019)

4.4 Trickle-Bed Reactors

Trickle-bed reactors (TBRs) are widely utilized in chemical and related industry sectors such as petroleum, petrochemical, coal industries, pharmaceuticals, and waste treatment. In certain wastewater treatment systems, trickle bed reactors are also effective and such reactors convert harmful phenol in wastewater into non-toxic CO_2 and H_2O (Azarpour et al., 2021; Feickert Fenske et al., 2023). TBR multiphase fluid dynamics are complicated, and they have a direct impact on ultimate reactor achievement in aspects of rate of reaction, product yield, and specificity. However, due to the complicated coupling among fluid dynamics and reaction mechanisms, traditional TBR modelling techniques cannot adequately account for such non-ideal behaviours. Recent developments in the implementation of CFD to three-phase TBR processes have shown the potential in accomplishing depth knowledge of the connections among multiphase hydrodynamics and chemical kinetics (Bouras et al., 2022; Wang et al., 2013).

TBRs are popularly used in wastewater treatment processes as a significant reactor design for gas/liquid/solid catalytic reactions. Catalytic wet air oxidation is a successful approach for treating wastewater, and the method was proved in a TBR utilizing a low-cost pillared clay catalyst at the lab scale (Tan et al., 2021). However, the complicated interaction of hydrodynamics and reaction mechanisms makes it extremely difficult to ramp up experimental reactors to advanced manufacturing or industrial reactors. Lopes et al. (2007) conducted hydrodynamics studies in a TBR in the catalytic wet oxidation of vanillic acid using a laboratory-made Mn-Ce–O catalyst. These researchers used a fluid dynamic model (CFD Eulerian model) for predicting pressure drop and fluid holdup for a TBR. The results show that the CFD Eulerian model anticipates fluid holdup quite well in the effective range of gas flow (0.10–0.70 kg/m^2s). Furthermore, the total organic carbon characteristics suggested that complete organic carbon reduction was accomplished at space times of up to 1.5 h.

The phenolic components are found in the pollutants of a wide range of industries, including oil refining, petrochemicals, pharmaceuticals, plastics, paint, and wood product manufacturing. These compounds are very harmful to many aquatic organisms (Makatsa et al., 2021). The catalytic wet air oxidation of phenolic in a TBR was modelled by using CFD at pressures ranging from 10 to 30 bars and temperatures ranging from 170 to 200 °C (Lopes & Quinta-Ferreira, 2007). These authors successfully created an Euler–Euler computational fluid model for the fluid dynamic estimation of a TBR designed for enhanced wastewater treatment infrastructure. According to the CFD results, fluid dynamic research shows that fluid holdup tends to increase as fluid flux rises and falls for elevated operating pressure levels. Furthermore, these researchers have found that phenolic compound conversion into a harmless product is strongly influenced by the temperature of the bed, whereas pressure has only a minimal impact on final conversion. In 2021 (Makatsa et al., 2021), similar CFD model was established from experimental data utilizing an Euler–Euler model to fully comprehend the actions of fluid flow within a TBR. The authors studied fluid flow behaviour, temperature profile, and the oxidation reactions of pillared clay catalyst containing TBR using the software package Fluent. According to the findings of this study, a hot spot was created close to the middle of the reactor as a consequence of fluid misdistribution. Furthermore, using a monolithic configuration in a reactor packing content reduced the pressure loss caused by reduced velocity within monolith channels. Moreover, at 160 °C and 10 bars, the phenol was totally oxidized to CO_2.

4.5 *Fluidized Bed Bioreactors*

Fluidized-bed bioreactors (FBRs) are versatile systems used in bioprocessing applications. They are solid particles suspended and fluidized by a continuous passage of gas or liquid. These reactors offer improved mass transfer, blending, and surface area for increased bioreaction efficiency. CFD has emerged as a highly appealing substitute to traditional experimental techniques in the field of fluid dynamics research. However, this numerical method needs a very effective and careful validation procedure. Eulerian–Eulerian and Eulerian–Lagrangian approaches are two approaches that are frequently used for the formulation of multifluid flow and motion of dispersed phase, respectively (Panneerselvam et al., 2007). CFD models for various systems, like solid–liquid, gas–solid and particle–fluid, have been used in the last two decades ago. The two-fluid model (TFM), based on the Eulerian–Eulerian approach, has been adopted to solve various multiphase flow dilemmas. The TFM requires a package of empirical or physical models for closing the conservation equations (Cornelissen et al., 2007). Nowadays, the kinetic theory of granular flow (KTGF) has become the top choice for the closure law to explain solid phase dynamics (solid viscosity, solid pressure, solid shear stress) along with the TFM (Xiao et al., 2023). However, some researchers considered a drag force model to obtain precise simulation results. Recently, some researchers applied the drag force model combined with the kinetic theory of the granular flow-based TFM model for the calculation of dynamic parameters and simulation hydrodynamics of gas–solid fluidized bed (Hosseini et al., 2009). Several momentum exchange coefficients in the gas–solid flow were also developed. Hosseini et al. (2013) examined different drag models at high gas velocities using a combined 2D Eulerian–Eulerian approach and the KTGF. They showed that particle motion and bubbles behaviour are satisfied by experimental data.

The CFD model incorporated both the TFM for the liquid particles two-phase fluidization system and the $c^2 - \varepsilon_c$ model for turbulent mass transfer. To characterize turbulence in the phases, the $kl - cl - kp - \varepsilon p - \Theta$ equations were utilized. Various simulations of hydrodynamics were performed using various drag models and modelling parameters to assess their sensitivity. The $c^2 - \varepsilon_c$ model formulations were used to precisely characterize turbulent mass diffusion in the liquid–solids fluidized bed (LSFB) during the adsorption process. The proposed model made it possible to acquire velocity and concentration fields. The simulation results and experimental data were compared and found to be in good agreement. The model demonstrated sensitivity to various drag models and excellent accord with experimental data. This provides important insights into the hydrodynamics and adsorption behaviour of LSFBs, allowing for the optimization of LSFB design and operation for protein adsorption processes (Ren et al., 2021). The study conducted by Chen et al. (2023) proposed a strategy for separating activated sludge selectively in a FBR using a CFD model (k-ε turbulence flow). It was discovered that the fluid field influences the distribution of sediment with various properties and functions. On the basis of floc size distribution, three components of sediment were identified. To evaluate the effect of selective sediment discharge on biotreatment efficiency, CFD simulations were performed. By selectively discharging decomposed sediment with the same energy input, the biotreatment efficiency could be increased by more than 5%, according to the results. This method demonstrated the feasibility of using the fluid field to identify and separate functionally decomposed sediment, resulting in improved bioreaction efficiency, energy utilization, and reactor stability in wastewater treatment.

However, the existing literature focuses primarily on gas–liquid or gas–solid interactions. Koerich et al. (2018) conducted a study to evaluate the effect of drag force, lift force, and particle collisions on the prediction of bioparticle fluidization. Modifications were made to the drag model to improve its accord with experimental data. Using a two-phase, transient, and turbulent approach on a symmetrical 3D, the researchers ran simulations. The results demonstrated a redistribution of bioparticles as a result of the lift force, but the restitution coefficient had a negligible effect on the fluidized bed height. In the case of bioparticles, the conventional drag models were insufficient for predicting the height of the liquid–solid fluidized bed. Nevertheless, modifying the drag force with a dimensionless number enhanced the agreement with experimental data across all operational conditions.

Furthermore, the key obstacles include accurately depicting complex multiphase flow behaviour, addressing the interaction between biological and hydrodynamic processes, and incorporating specific biokinetic and biochemical reactions. Obtaining reliable experimental data for validation remains a further obstacle. Future directions for CFD modelling in FBR include refining turbulence models, enhancing formulations for drag and lift forces, and developing more precise models for heat and mass transfer. Integration of CFD with computational biology and multiscale modelling techniques can enhance understanding of reactor performance. In addition, advancements in high-performance computing and increased access to experimental data can facilitate the development of more robust and dependable CFD models for FBRs.

5 Future Prospective

There are some of the challenges in CFD and bioreactor modelling which include capturing complex multiphase flow behaviour, accurately representing biological processes, addressing fouling and membrane interactions, validating models with limited experimental data, and optimizing computational resources for large simulations.

The application of CFD in activated sludge bioreactors encounters challenges related to modelling biological and hydrodynamic interactions, representing non-Newtonian behaviour, handling diverse particle sizes, and predicting biomass settling. However, these challenges can be addressed by integrating population balance models, enhancing biokinetic and biochemical models, developing advanced simulation tools, and accurately representing microbial interactions (Karpinska & Bridgeman, 2016a; Samstag et al., 2016a). Similarly, in fixed bed reactor, capturing complex flow patterns, predicting pressure drop distribution, modelling heat and mass transfer in porous media, and representing reaction kinetics present significant obstacles that can be overcome by employing advanced techniques for modelling porous media, incorporating detailed reaction kinetics, coupling with heat transfer models, and utilizing multiscale modelling approaches to capture catalyst deactivation (Malang et al., 2015).

In the context of MBR, it is difficult to predict membrane fouling, model multiphase flow and mass transfer, comprehend biological-filtration interactions, and capture complex flow patterns. Future directions, however, aim to address these obstacles by developing advanced fouling models, enhancing the representation of multiphase flow, incorporating biofilm models, optimizing membrane module design, and coupling these models with advanced filtration approaches (Jalilnejad et al., 2022; Shi et al., 2021). Similarly, modelling gas–liquid–solid interactions, accurately predicting liquid distribution and pressure drop, and representing reaction kinetics are obstacles for TBRs. Future efforts will concentrate on the development of advanced liquid distributor models, the improvement of mass transfer modelling, the incorporation of detailed reaction kinetics, the consideration of particle dynamics, and the optimization of reactor design for enhanced mass transfer efficiency (Azarpour et al., 2021; Wang et al., 2013). In FBR, the difficulties include the modelling of multiphase flow, the accurate prediction of hydrodynamics and particle–fluid interactions, the comprehension of bubble distribution and solids mixing, and the maintenance of reactor stability. Future directions consist of advancing numerical techniques for fluidization dynamics, enhancing particle–fluid interaction models, incorporating detailed biochemical models, optimizing reactor design to improve mixing and reaction efficiency, and developing control strategies to enhance overall stability (Lettieri & Mazzei, 2009; Pan et al., 2016).

6 Conclusion

For efficient and sustainable bioprocesses, the bioreactor development needs to tackle some of the challenges such as optimization of mass transfer and mixing, controlling microbial behaviour and product formation, managing fouling and biofilm formation, ensuring scalability and reproducibility, and integrating advanced monitoring and control systems. CFD modelling has proven to be an effective tool for understanding and optimizing different types of bioreactors. Various CFD models such as the Eulerian–Eulerian model, Eulerian–Lagrangian model, two-fluid model, population balance model, biofilm models, porous media models, etc. have been used to simulate the complex flow patterns, hydrodynamics, mass transfer, and fouling phenomena that occur in bioreactors. These models shed light on the characteristics of multiphase flows, including gas–liquid–solid interactions, bubble distribution, particle–fluid dynamics, and biofilm formation. By combining CFD models with experimental data and sophisticated simulation techniques, bioreactor designs can be optimized, operational parameters can be fine-tuned, and overall process efficiency can be enhanced. Research and development in CFD modelling for bioreactors holds great promise for advancing bioprocess engineering and enhancing the performance of various bioreactor systems.

References

Abali, B. E. (2019). An accurate finite element method for the numerical solution of isothermal and incompressible flow of viscous fluid. *Fluids, 4*(1), 5. https://doi.org/10.3390/fluids4010005

Afshari, F., Zavaragh, H. G., Sahin, B., Grifoni, R. C., Corvaro, F., Marchetti, B., & Polonara, F. (2018). On numerical methods; optimization of CFD solution to evaluate fluid flow around a sample object at low Re numbers. *Mathematics and Computers in Simulation, 152*, 51–68. https://doi.org/10.1016/j.matcom.2018.04.004

Ahmad, T., & Zhang, D. (2020). A critical review of comparative global historical energy consumption and future demand: The story told so far. *Energy Reports, 6*, 1973–1991. https://doi.org/10.1016/j.egyr.2020.07.020

Ahmmed, M. S., Skerman, A., & Batstone, D. J. (2022). Predicting long-term solid accumulation in waste stabilisation lagoons through a combined CFD-process model approach. *Chemical Engineering Research and Design, 184*, 267–276. https://doi.org/10.1016/j.cherd.2022.06.012

Alidrisi, H., & Demirbas, A. (2016). Enhanced electricity generation using biomass materials. *Energy Sources, Part a: Recovery, Utilization, and Environmental Effects, 38*(10), 1419–1427. https://doi.org/10.1080/15567036.2014.948647

Alobaid, F., Almohammed, N., Massoudi Farid, M., May, J., Rößger, P., Richter, A., & Epple, B. (2022). Progress in CFD Simulations of Fluidized Beds for Chemical and Energy Process Engineering. *Progress in Energy and Combustion Science, 91*, 100930. https://doi.org/10.1016/j.pecs.2021.100930

Aly, H. S., Eldrainy, Y. A., Saqr, K. M., Lazim, T. M., & Jaafar, M. N. M. (2010). A mathematical model for predicting spray atomization characteristics in an Eulerian-Eulerian framework. *International Communications in Heat and Mass Transfer, 37*(6), 618–623. https://doi.org/10.1016/j.icheatmasstransfer.2010.02.003

Amato, T., & Wicks, J. (2009). *Dissolved air flotation and potential clarified water quality based on computational fluid dynamics modelling CFD with reactions-simulation of wastewater treatment plants view project*. https://doi.org/10.13140/2.1.2605.2480

Arif, A. U. A., Sorour, M. T., & Aly, S. A. (2020). Cost analysis of activated sludge and membrane bioreactor WWTPs using CapdetWorks simulation program: Case study of Tikrit WWTP (middle Iraq). *Alexandria Engineering Journal, 59*(6), 4659–4667. https://doi.org/10.1016/j.aej.2020.08.023

Arumuga Perumal, D., & Dass, A. K. (2015). A Review on the development of lattice Boltzmann computation of macro fluid flows and heat transfer. *Alexandria Engineering Journal, 54*(4), 955–971. https://doi.org/10.1016/j.aej.2015.07.015

Azarpour, A., Rezaei, N., & Zendehboudi, S. (2021). Performance analysis and modeling of catalytic trickle-bed reactors: A comprehensive review. *Journal of Industrial and Engineering Chemistry, 103*, 1–41. https://doi.org/10.1016/j.jiec.2021.04.020

Bach, C., Yang, J., Larsson, H., Stocks, S. M., Gernaey, K. V., Albaek, M. O., & Krühne, U. (2017). Evaluation of mixing and mass transfer in a stirred pilot scale bioreactor utilizing CFD. *Chemical Engineering Science, 171*, 19–26. https://doi.org/10.1016/j.ces.2017.05.001

Balakin, B. V., Hoffmann, A. C., Kosinski, P., & Rhyne, L. D. (2010). Eulerian-Eulerian CFD model for the sedimentation of spherical particles in suspension with high particle concentrations. *Engineering Applications of Computational Fluid Mechanics, 4*(1), 116–126. https://doi.org/10.1080/19942060.2010.11015303

Basavarajappa, M., & Miskovic, S. (2016). Investigation of gas dispersion characteristics in stirred tank and flotation cell using a corrected CFD-PBM quadrature-based moment method approach. *Minerals Engineering, 95*, 161–184. https://doi.org/10.1016/j.mineng.2016.06.026

Batstone, D. J., Puyol, D., Flores-Alsina, X., & Rodríguez, J. (2015). Mathematical modelling of anaerobic digestion processes: Applications and future needs. *Reviews in Environmental Science and Biotechnology, 14*(4), 595–613. https://doi.org/10.1007/s11157-015-9376-4

Behin, J., & Bahrami, S. (2012). Modeling an industrial dissolved air flotation tank used for separating oil from wastewater. *Chemical Engineering and Processing: Process Intensification, 59*, 1–8. https://doi.org/10.1016/j.cep.2012.05.004

Bhanduvula, S. (2012). Finite difference method in computational fluid dynamics. *International Journal of Education and Applied Research, 2*.

Bhaskaran, R., & Collins, L. (2020). *Introduction to CFD basics*.

Bhatti, M. M., Marin, M., Zeeshan, A., & Abdelsalam, S. I. (2020). Editorial: Recent trends in computational fluid dynamics. *Frontiers in Physics*. https://doi.org/10.3389/fphy.2020.593111

Bhuyar, L. B., Thakre, S. B., & Ingole, N. W. (2009). Design characteristics of Curved Blade Aerator w.r.t. aeration efficiency and overall oxygen transfer coefficient and comparison with CFD modeling. *International Journal of Engineering, Science and Technology, 1*(1). www.ijest-ng.com

Boas, J. V., Oliveira, V. B., Simões, M., & Pinto, A. M. F. R. (2022). Review on microbial fuel cells applications, developments and costs. *Journal of Environmental Management, 307*, 114525. https://doi.org/10.1016/j.jenvman.2022.114525

Bouras, H., Haroun, Y., Fortunato Bodziony, F., Philippe, R., Fongarland, P., & Augier, F. (2022). Use of CFD for pressure drop, liquid saturation and wetting predictions in trickle bed reactors for different catalyst particle shapes. *Chemical Engineering Science, 249*, 117315. https://doi.org/10.1016/j.ces.2021.117315

Brannock, M., Wang, Y., & Leslie, G. (2010). Mixing characterisation of full-scale membrane bioreactors: CFD modelling with experimental validation. *Water Research, 44*(10), 3181–3191. https://doi.org/10.1016/j.watres.2010.02.029

Camp, T. R. (1942). Grit Chamber Design. *Source: Sewage Works Journal, 14*(2). https://about.jstor.org/terms

Cantarero Rivera, F. J., & Chen, J. (2022). Computational fluid dynamics modeling of cell cultures in bioreactors and its potential for cultivated meat production—A mini-review. *Future Foods, 6*, 100195. https://doi.org/10.1016/j.fufo.2022.100195

Carey, R. O., & Migliaccio, K. W. (2009). Contribution of wastewater treatment plant effluents to nutrient dynamics in aquatic systems. *Environmental Management*, 44(2), 205–217. https://doi.org/10.1007/s00267-009-9309-5

Carpentieri, M., Kumar, P., & Robins, A. (2011). An overview of experimental results and dispersion modelling of nanoparticles in the wake of moving vehicles. *Environmental Pollution, 159*(3), 685–693. https://doi.org/10.1016/j.envpol.2010.11.041

Casula, E., Molognoni, D., Borràs, E., & Mascia, M. (2021). 3D modelling of bioelectrochemical systems with brush anodes under fed-batch and flow conditions. *Journal of Power Sources, 487*, 229432. https://doi.org/10.1016/j.jpowsour.2020.229432

Chang, J., Zhang, K., Chen, H., Yang, Y., & Zhang, L. (2013). CFD modelling of the hydrodynamics and kinetic reactions in a fluidised-bed MTO reactor. *Chemical Engineering Research and Design, 91*(12), 2355–2368. https://doi.org/10.1016/J.CHERD.2013.04.023

Chen, B., Wei, G., Zhang, T., Wu, H., Wu, C., Chen, A., Zhang, H., Guan, X., Ren, Y., Feng, C., Qiu, G., & Wei, C. (2023). Flow drag force contributes high bio-treatment efficiency in a circulating fluidized bed reactor: Mechanism of selective separation of functional decayed sludge. *Chemical Engineering Journal, 454*, 140448. https://doi.org/10.1016/j.cej.2022.140448

Climent, J., Basiero, L., Martínez-Cuenca, R., Berlanga, J. G., Julián-López, B., & Chiva, S. (2018). Biological reactor retrofitting using CFD-ASM modelling. *Chemical Engineering Journal, 348*, 1–14. https://doi.org/10.1016/j.cej.2018.04.058

Cornelissen, J. T., Taghipour, F., Escudié, R., Ellis, N., & Grace, J. R. (2007). CFD modelling of a liquid–solid fluidized bed. *Chemical Engineering Science, 62*(22), 6334–6348. https://doi.org/10.1016/j.ces.2007.07.014

D'Bastiani, C., Kennedy, D., & Reynolds, A. (2023). CFD simulation of anaerobic granular sludge reactors: A review. *Water Research, 242*, 120220. https://doi.org/10.1016/j.watres.2023.120220

Dabiri, S., Kumar, P., & Rauch, W. (2023). Integrating biokinetics with computational fluid dynamics for energy performance analysis in anaerobic digestion. *Bioresource Technology, 373*, 128728. https://doi.org/10.1016/j.biortech.2023.128728

Dai, Z., Yu, R., Wu, Y., Zhu, G., Lu, X., & Zha, X. (2022). Modelling of self-sustainable microbial fuel cell type oil sensors based on restricted oxygen transfer and two-population competition. *Science of the Total Environment, 806*, 151333. https://doi.org/10.1016/j.scitotenv.2021.151333

Das, S., Bai, H., Wu, C., Kao, J. H., Barney, B., Kidd, M., & Kuettel, M. (2016). Improving the performance of industrial clarifiers using three-dimensional computational fluid dynamics. *Engineering Applications of Computational Fluid Mechanics, 10*(1), 130–144. https://doi.org/10.1080/19942060.2015.1121518

Deb, D., Patel, R., & Balas, V. E. (2020). A Review of Control-Oriented Bioelectrochemical Mathematical Models of Microbial Fuel Cells. *Processes, 8*(5), 583. https://doi.org/10.3390/pr8050583

Dehnavi, M. A., Shahhosseini, S., Hashemabadi, S. H., & Ghafelebashi, S. M. (2010). CFD simulation of hydrodynamics and heat transfer in gas phase ethylene polymerization reactors. *International Communications in Heat and Mass Transfer, 37*(4), 437–442. https://doi.org/10.1016/j.icheatmasstransfer.2009.12.005

Dick, E. (n.d.). Introduction to finite volume methods in computational fluid dynamics. In *Computational fluid dynamics* (pp. 275–301). Springer, Berlin. https://doi.org/10.1007/978-3-540-85056-4_11

Dick, E. (2009). Introduction to finite element methods in computational fluid dynamics. In *Computational fluid dynamics* (pp. 235–274). Springer, Berlin. https://doi.org/10.1007/978-3-540-85056-4_10

Dixon, A. G., & Partopour, B. (2020). Computational Fluid Dynamics for Fixed Bed Reactor Design. *Annual Review of Chemical and Biomolecular Engineering, 11*(1), 109–130. https://doi.org/10.1146/annurev-chembioeng-092319-075328

Dong, Y., Sosna, B., Korup, O., Rosowski, F., & Horn, R. (2017). Investigation of radial heat transfer in a fixed-bed reactor: CFD simulations and profile measurements. *Chemical Engineering Journal, 317*, 204–214. https://doi.org/10.1016/j.cej.2017.02.063

Doumic, L. I., Soares, P. A., Ayude, M. A., Cassanello, M., Boaventura, R. A. R., & Vilar, V. J. P. (2015). Enhancement of a solar photo-Fenton reaction by using ferrioxalate complexes for the treatment of a synthetic cotton-textile dyeing wastewater. *Chemical Engineering Journal, 277*, 86–96. https://doi.org/10.1016/j.cej.2015.04.074

Dutta, A., Das, N., Sarkar, D., & Chakrabarti, S. (2019). Development and characterization of a continuous solar-collector-reactor for wastewater treatment by photo-Fenton process. *Solar Energy, 177*, 364–373. https://doi.org/10.1016/j.solener.2018.11.036

Ed Fontes. (2018, November 29). FEM vs. FVM. COMSOL.

Eshtiaghi, N., Markis, F., Yap, S. D., Baudez, J. C., & Slatter, P. (2013). Rheological characterisation of municipal sludge: A review. In *Water Research* (Vol. 47, Issue 15, pp. 5493–5510). Elsevier Ltd. https://doi.org/10.1016/j.watres.2013.07.001

Falcão, D. M. (1997). *High performance computing in power system applications* (pp. 1–23). https://doi.org/10.1007/3-540-62828-2_109

Farber, P., Gräbel, J., Kroppen, N., Pötschke, L., Roos, D., Rosenbaum, M., Stegschuster, G., & Ueberholz, P. (2021). Electricity generation in a microbial fuel cell with textile carbon fibre anodes. *Computers & Mathematics with Applications, 83*, 4–23. https://doi.org/10.1016/j.camwa.2019.11.019

Farzan, P., Mistry, B., & Ierapetritou, M. G. (2017). Review of the important challenges and opportunities related to modeling of mammalian cell bioreactors. *AIChE Journal, 63*(2), 398–408. https://doi.org/10.1002/aic.15442

Feickert Fenske, C., Strübing, D., & Koch, K. (2023). Biological methanation in trickle bed reactors—a critical review. *Bioresource Technology, 385*, 129383. https://doi.org/10.1016/j.biortech.2023.129383

Fu, B., Xu, T., Guo, X., Liang, P., Huang, X., & Zhang, X. (2019). Optimization and simulation of a carbon-based flow-through composite anode configuration to enhance power generation and improve effluent quality simultaneously for microbial fuel cells. *Journal of Cleaner Production, 229*, 542–551. https://doi.org/10.1016/j.jclepro.2019.04.308

Gadkari, S., Gu, S., & Sadhukhan, J. (2018). Towards automated design of bioelectrochemical systems: A comprehensive review of mathematical models. *Chemical Engineering Journal, 343*, 303–316. https://doi.org/10.1016/j.cej.2018.03.005

Gao, H., & Stenstrom, M. K. (n.d.). *World environmental and water resources congress 2020.*

Gao, H., & Stenstrom, M. K. (2020). Development and applications in computational fluid dynamics modeling for secondary settling tanks over the last three decades: A review. In *Water environment research* (Vol. 92, Issue 6, pp. 796–820). Wiley, New York. https://doi.org/10.1002/wer.1279

Glessmer, M., & Janßen, C. (2017). Using an interactive Lattice Boltzmann solver in fluid mechanics instruction. *Computation, 5*(4), 35. https://doi.org/10.3390/computation5030035

Gómez-Ríos, D., Junne, S., Neubauer, P., Ochoa, S., Ríos-Estepa, R., & Ramírez-Malule, H. (2019). Characterization of the metabolic response of Streptomyces clavuligerus to shear stress in stirred tanks and single-use 2D rocking motion bioreactors for clavulanic acid production. *Antibiotics, 8*(4), 168. https://doi.org/10.3390/antibiotics8040168

Goswami, L., Vinoth Kumar, R., Borah, S. N., Arul Manikandan, N., Pakshirajan, K., & Pugazhenthi, G. (2018). Membrane bioreactor and integrated membrane bioreactor systems for micropollutant removal from wastewater: A review. *Journal of Water Process Engineering, 26*, 314–328. https://doi.org/10.1016/j.jwpe.2018.10.024

Gu, Y., Li, Y., Yuan, F., & Yang, Q. (2023). Optimization and control strategies of aeration in WWTPs: A review. *Journal of Cleaner Production.* https://doi.org/10.1016/j.jclepro.2023.138008

Guzmán, C., Carrera, J., Durán, H., Berumen, J., Ortiz, A., Guirette, O., Arroyo, A., Brizuela, J., Gómez, F., Blanco, A., Azcaray, H., & Hernández, M. (2018). Implementation of virtual sensors for monitoring temperature in greenhouses using CFD and control. *Sensors, 19*(1), 60. https://doi.org/10.3390/s19010060

Hassanzadeh, A., Firouzi, M., Albijanic, B., & Celik, M. S. (2018). A review on determination of particle–bubble encounter using analytical, experimental and numerical methods. *Minerals Engineering, 122*, 296–311. https://doi.org/10.1016/j.mineng.2018.04.014

He, W., Li, Z. K., Tang, A. R., Liu, A., & He, B. G. (2013). Simulation study of three-phase flow field based on microbubble flotation. *Applied Mechanics and Materials, 477–478*, 226–231. https://doi.org/10.4028/www.scientific.net/AMM.477-478.226

Heroux, M. A. (2022). Research software science: Expanding the impact of research software engineering. *Computing in Science & Engineering, 24*(6), 22–27. https://doi.org/10.1109/MCSE.2023.3260475

Hickman, D. A., Degenstein, J. C., & Ribeiro, F. H. (2016). Fundamental principles of laboratory fixed bed reactor design. *Current Opinion in Chemical Engineering, 13*, 1–9. https://doi.org/10.1016/j.coche.2016.07.002

Hoiberg, B., & Shah, M. T. (2021). CFD study of multiphase flow in aerated grit tank. *Journal of Water Process Engineering*. https://doi.org/10.1016/j.jwpe.2020.101698

Hosseini, S. H., Ahmadi, G., & Olazar, M. (2013). CFD simulation of cylindrical spouted beds by the kinetic theory of granular flow. *Powder Technology, 246*, 303–316. https://doi.org/10.1016/j.powtec.2013.05.016

Hosseini, S. H., Rahimi, R., Zivdar, M., & Samimi, A. (2009). The effect of ring baffles on the hydrodynamics of a gas—Solid bubbling fluidized bed using computational fluid dynamics. *Proceedings of the Institution of Mechanical Engineers, Part c: Journal of Mechanical Engineering Science, 223*(10), 2281–2289. https://doi.org/10.1243/09544062JMES1575

Hreiz, R., Potier, O., Wicks, J., & Commenge, J. M. (2019). CFD Investigation of the effects of bubble aerator layouts on hydrodynamics of an activated sludge channel reactor. *Environmental Technology (united Kingdom), 40*(20), 2657–2670. https://doi.org/10.1080/09593330.2018.1448001

https://www.cfdsupport.com/hardware-for-cfd.html. (n.d.). *Hardware for CFD simulations.*

https://www.comsol.com/. (n.d.). *COMSOL Multiphysics® simulation software understand, predict, and optimize real-world designs, devices, and processes with simulation.*

Hyre, M. (2012). *Evaluation of three full-scale grit removal processes using CFD modeling.* https://www.researchgate.net/publication/265786513

Itoh, N., & Mimura, K. (2013). Computational fluid dynamics (CFD) analysis of membrane reactors: Simulation of single-and multi-tube palladium membrane reactors for hydrogen recovery from cyclohexane. In *Handbook of Membrane Reactors* (Vol. 1, pp. 464–495). Elsevier Inc. https://doi.org/10.1533/9780857097330.3.464

Jafarinejad, S. (2017). Cost estimation and economical evaluation of three configurations of activated sludge process for a wastewater treatment plant (WWTP) using simulation. *Applied Water Science, 7*(5), 2513–2521. https://doi.org/10.1007/S13201-016-0446-8

Jalilnejad, E., Jabbari, B., & Ghasemzadeh, K. (2022). Application of computational fluid dynamics technique in membrane bioreactor systems. In *Current trends and future developments on (Bio-) membranes* (pp. 345–375). Elsevier. https://doi.org/10.1016/B978-0-12-822294-2.00007-2

Jamshidian, R., Scully, J., & Van den Akker, H. E. A. (2023). Two-fluid simulations of an aerated lab-scale bioreactor. *Chemical Engineering Research and Design, 196*, 254–275. https://doi.org/10.1016/j.cherd.2023.06.042

Jasak, H. (2009). OpenFOAM: Open source CFD in research and industry. *International Journal of Naval Architecture and Ocean Engineering, 1*(2), 89–94. https://doi.org/10.2478/IJNAOE-2013-0011

Jasak, H., & Tuković. (2010). *Dynamic mesh handling in openfoam applied to fluid-structure interaction simulations.*

Jasak, H., & Uroić, T. (2020). *Practical computational fluid dynamics with the finite volume method* (pp. 103–161). CISM International Centre for Mechanical Sciences. https://doi.org/10.1007/978-3-030-37518-8_4

Jeong, W., & Seong, J. (2014). Comparison of effects on technical variances of computational fluid dynamics (CFD) software based on finite element and finite volume methods. *International Journal of Mechanical Sciences, 78*, 19–26. https://doi.org/10.1016/j.ijmecsci.2013.10.017

Jin, Y., Liu, C. L., Song, X. F., & Yu, J. G. (2019). Computational fluid dynamics simulation as a tool for optimizing the hydrodynamic performance of membrane bioreactors. *RSC Advances, 9*(55), 32034–32046. https://doi.org/10.1039/c9ra06706j

John, D. (n.d.). *Improved design of settling tanks using an extended drift flux model.* http://research-information.bristol.ac.uk

Jurtz, N., Kraume, M., & Wehinger, G. D. (2019). Advances in fixed-bed reactor modeling using particle-resolved computational fluid dynamics (CFD). *Reviews in Chemical Engineering, 35*(2), 139–190. https://doi.org/10.1515/revce-2017-0059

Kamyar, A., Saidur, R., & Hasanuzzaman, M. (2012). Application of computational fluid dynamics (CFD) for nanofluids. *International Journal of Heat and Mass Transfer, 55*(15–16), 4104–4115. https://doi.org/10.1016/j.ijheatmasstransfer.2012.03.052

Karn, A. L., Pandya, S., Mehbodniya, A., Arslan, F., Sharma, D. K., Phasinam, K., Aftab, M. N., Rajan, R., Bommisetti, R. K., & Sengan, S. (2023). An integrated approach for sustainable development of wastewater treatment and management system using IoT in smart cities. *Soft Computing, 27*(8), 5159–5175. https://doi.org/10.1007/s00500-021-06244-9

Karpinska, A. M., & Bridgeman, J. (2016a). CFD-aided modelling of activated sludge systems—A critical review. *Water Research, 88*, 861–879. https://doi.org/10.1016/j.watres.2015.11.008

Karthik, G. M., & Buwa, V. V. (2018). Particle-resolved simulations of methane steam reforming in multilayered packed beds. *AIChE Journal, 64*(11), 4162–4176. https://doi.org/10.1002/aic.16386

Kim, J., Kim, H., Kim, B., & Yu, J. (2014). Computational fluid dynamics analysis in microbial fuel cells with different anode configurations. *Water Science and Technology, 69*(7), 1447–1452. https://doi.org/10.2166/wst.2014.041

Koerich, D. M., Lopes, G. C., & Rosa, L. M. (2018). Investigation of phases interactions and modification of drag models for liquid-solid fluidized bed tapered bioreactors. *Powder Technology, 339*, 90–101. https://doi.org/10.1016/j.powtec.2018.07.102

Kone, J.-P., Zhang, X., Yan, Y., Hu, G., & Ahmadi, G. (2017). Three-dimensional multiphase flow computational fluid dynamics models for proton exchange membrane fuel cell: A theoretical development. *The Journal of Computational Multiphase Flows, 9*(1), 3–25. https://doi.org/10.1177/1757482X17692341

Korzani, M. G., Galindo Torres, S., Scheuermann, A., & Williams, D. J. (2016). Smoothed particle hydrodynamics into the fluid dynamics of classical problems. *Applied Mechanics and Materials, 846*, 73–78. https://doi.org/10.4028/www.scientific.net/AMM.846.73

Kostoglou, M., Karapantsios, T. D., & Matis, K. A. (2007). CFD model for the design of large scale flotation tanks for water and wastewater treatment. *Industrial and Engineering Chemistry Research, 46*(20), 6590–6599. https://doi.org/10.1021/ie0703989

Kumar, T., & Eswari, J. S. (2023). Review and perspectives of emerging green technology for the sequestration of carbon dioxide into value-added products: An intensifying development. *Energy & Fuels, 37*(5), 3570–3589. https://doi.org/10.1021/acs.energyfuels.2c04122

Kumar, T., & Eswari Jujjavarappu, S. (2023). A critical review on an advanced bio-electrochemical system for carbon dioxide sequestration and wastewater treatment. *Total Environment Research Themes, 5*, 100023. https://doi.org/10.1016/j.totert.2022.100023

Kumar, T., & Jujjavarapu, S. E. (2023a). Advances in anode configurations for a microbial fuel cell via a computational fluid dynamics electrochemistry and its experimental validation. *Journal of Chemical Technology & Biotechnology*. https://doi.org/10.1002/jctb.7397

Kumar, T., & Jujjavarapu, S. E. (2023b). Carbon dioxide sequestration and wastewater treatment via an innovative self-sustaining algal microbial fuel cell. *Journal of Cleaner Production, 415*, 137836. https://doi.org/10.1016/j.jclepro.2023.137836

Kumar, T., Naik, S., & Jujjavarappu, S. E. (2021). A critical review on early-warning electrochemical system on microbial fuel cell-based biosensor for on-site water quality monitoring. *Chemosphere*. https://doi.org/10.1016/j.chemosphere.2021.133098

Kumar, T., Naik, S., & Jujjavarappu, S. E. (2022). A critical review on early-warning electrochemical system on microbial fuel cell-based biosensor for on-site water quality monitoring. *Chemosphere, 291*, 133098. https://doi.org/10.1016/j.chemosphere.2021.133098

Kwon, S. B., Park, N. S., Lee, S. J., Ahn, H. W., & Wang, C. K. (2006). Examining the effect of length/width ratio on the hydro-dynamic behaviour in a DAF system using CFD and ADV techniques. *Water Science and Technology, 53*(7), 141–149. https://doi.org/10.2166/wst.2006.218

Lamon, L., Asturiol, D., Vilchez, A., Cabellos, J., Damásio, J., Janer, G., Richarz, A., & Worth, A. (2019). Physiologically based mathematical models of nanomaterials for regulatory toxicology: A review. *Computational Toxicology, 9*, 133–142. https://doi.org/10.1016/j.comtox.2018.10.002

le Moullec, Y., Gentric, C., Potier, O., & Leclerc, J. P. (2010). CFD simulation of the hydrodynamics and reactions in an activated sludge channel reactor of wastewater treatment. *Chemical Engineering Science, 65*(1), 492–498. https://doi.org/10.1016/j.ces.2009.03.021

Lee, I.-B., Bitog, J. P. P., Hong, S.-W., Seo, I.-H., Kwon, K.-S., Bartzanas, T., & Kacira, M. (2013). The past, present and future of CFD for agro-environmental applications. *Computers and Electronics in Agriculture, 93*, 168–183. https://doi.org/10.1016/j.compag.2012.09.006

Lee, K. H., Kim, H., KuK, J. W., Chung, J. D., Park, S., & Kwon, E. E. (2020). Micro-bubble flow simulation of dissolved air flotation process for water treatment using computational fluid dynamics technique. *Environmental Pollution*. https://doi.org/10.1016/j.envpol.2019.01.011

Lee, J. L., & Lim, E. W. C. (2017). Comparisons of Eulerian-Eulerian and CFD-DEM simulations of mixing behaviors in bubbling fluidized beds. *Powder Technology, 318*, 193–205. https://doi.org/10.1016/j.powtec.2017.05.050

Lettieri, P., & Mazzei, L. (2009). Challenges and Issues on the CFD modeling of fluidized beds: A review. *The Journal of Computational Multiphase Flows, 1*(2), 83–131. https://doi.org/10.1260/175748209789563937

Li, L., Wang, K., Zhao, Q., Gao, Q., Zhou, H., Jiang, J., & Mei, W. (2022). A critical review of experimental and CFD techniques to characterize the mixing performance of anaerobic digesters for biogas production. *Reviews in Environmental Science and Bio/technology, 21*(3), 665–689. https://doi.org/10.1007/s11157-022-09626-z

Li, W., Li, M., Bolton, J. R., & Qiang, Z. (2016). Configuration optimization of UV reactors for water disinfection with computational fluid dynamics: Feasibility of using particle minimum UV dose as a performance indicator. *Chemical Engineering Journal, 306*, 1–8. https://doi.org/10.1016/j.cej.2016.07.042

Li, W., Li, M., Wen, D., & Qiang, Z. (2018a). Development of economical-running strategy for multi-lamp UV disinfection reactors in secondary water supply systems with computational fluid dynamics simulations. *Chemical Engineering Journal, 343*, 317–323. https://doi.org/10.1016/j.cej.2018.03.017

Li, X., Scott, K., Kelly, W. J., & Huang, Z. (2018b). Development of a computational fluid dynamics model for scaling-up Ambr bioreactors. *Biotechnology and Bioprocess Engineering, 23*(6), 710–725. https://doi.org/10.1007/s12257-018-0063-5

Lim, S. S., Fontmorin, J.-M., Pham, H. T., Milner, E., Abdul, P. M., Scott, K., Head, I., & Yu, E. H. (2021). Zinc removal and recovery from industrial wastewater with a microbial fuel cell: Experimental investigation and theoretical prediction. *Science of the Total Environment, 776*, 145934. https://doi.org/10.1016/j.scitotenv.2021.145934

Lin, J.-R., Cao, J., Zhang, J.-P., van Treeck, C., & Frisch, J. (2019). Visualization of indoor thermal environment on mobile devices based on augmented reality and computational fluid dynamics. *Automation in Construction, 103*, 26–40. https://doi.org/10.1016/j.autcon.2019.02.007

Liu, X., Wang, Y., Shi, Y., Li, Q., Dai, P., Guan, J., Waite, T. D., & Leslie, G. (2019). CFD modelling of uneven flows behaviour in flat-sheet membrane bioreactors: From bubble generation to shear stress distribution. *Journal of Membrane Science, 570–571*, 146–155. https://doi.org/10.1016/j.memsci.2018.10.040

Liu, X., Wang, Y., Waite, T. D., & Leslie, G. (2015). Numerical simulation of bubble induced shear in membrane bioreactors: Effects of mixed liquor rheology and membrane configuration. *Water Research, 75*, 131–145. https://doi.org/10.1016/j.watres.2015.02.009

Llano-Serna, C. E., Coral-Portillo, J. A., Fontalvo, J., & Prado-Rubio, O. A. (2019). CFD assisted analysis and design of hydraulic flocculators. *Revista Mexicana de Ingeniera Quimica, 18*(3), 995–1015. https://doi.org/10.24275/UAM/IZT/DCBI/REVMEXINGQUIM/2019V18N3/LLANO

Logtenberg, S. A., & Dixon, A. G. (1998). Computational fluid dynamics studies of fixed bed heat transfer. *Chemical Engineering and Processing: Process Intensification, 37*(1), 7–21. https://doi.org/10.1016/S0255-2701(97)00032-9

Lopes, R. J. G., & Quinta-Ferreira, R. M. (2007). Trickle-bed CFD studies in the catalytic wet oxidation of phenolic acids. *Chemical Engineering Science, 62*(24), 7045–7052. https://doi.org/10.1016/j.ces.2007.08.085

Lopes, R. J. G., Silva, A. M. T., & Quinta-Ferreira, R. M. (2007). Kinetic modeling and trickle-bed CFD studies in the catalytic wet oxidation of vanillic acid. *Industrial & Engineering Chemistry Research, 46*(25), 8380–8387. https://doi.org/10.1021/ie070009a

López-Jiménez, P. A., Escudero-González, J., Montoya Martínez, T., Fajardo Montañana, V., & Gualtieri, C. (2015). Application of CFD methods to an anaerobic digester: The case of Ontinyent WWTP, Valencia, Spain. *Journal of Water Process Engineering, 7*, 131–140. https://doi.org/10.1016/j.jwpe.2015.05.006

Luo, S., Sun, H., Ping, Q., Jin, R., & He, Z. (2016). A review of modeling bioelectrochemical systems: Engineering and statistical aspects. *Energies, 9*(2), 111. https://doi.org/10.3390/en9020111

Macfarlan, L. H., Phan, M. T., & Bruce Eldridge, R. (2021). Structured packing geometry study for liquid-phase mass transfer and hydrodynamic performance using CFD. *Chemical Engineering Science*. https://doi.org/10.1016/J.CES.2021.117353

Makatsa, T. J., Baloyi, J., Ntho, T., & Masuku, C. M. (2021). Catalytic wet air oxidation of phenol: Review of the reaction mechanism, kinetics, and CFD modeling. *Critical Reviews in Environmental Science and Technology, 51*(17), 1891–1923. https://doi.org/10.1080/10643389.2020.1771886

Makatsa, T. J., Baloyi, S. J., & Masuku, C. M. (2021). *Computational fluid dynamics modelling of phenol oxidation in a trickle-bed reactor using 3D Eulerian model* (pp. 825–831). https://doi.org/10.1016/B978-0-323-88506-5.50129-7

Malang, J., Kumar, P., & Saptoro, A. (2015). Computational fluid dynamics-based hydrodynamics studies in packed bed columns: Current status and future directions. *International Journal of Chemical Reactor Engineering, 13*(3), 289–303. https://doi.org/10.1515/ijcre-2014-0121

Malato, S., Fernández-Ibáñez, P., Maldonado, M. I., Blanco, J., & Gernjak, W. (2009). Decontamination and disinfection of water by solar photocatalysis: Recent overview and trends. In *Catalysis today* (Vol. 147, Issue 1, pp. 1–59). https://doi.org/10.1016/j.cattod.2009.06.018

Mandenius, C.-F. (2016). Challenges for bioreactor design and operation. In *Bioreactors* (pp. 1–34). Wiley-VCH Verlag GmbH & Co. KGaA. https://doi.org/10.1002/9783527683369.ch1

Massaglia, G., Gerosa, M., Agostino, V., Cingolani, A., Sacco, A., Saracco, G., Margaria, V., & Quaglio, M. (2017). Fluid dynamic modeling for microbial fuel cell based biosensor optimization. *Fuel Cells, 17*(5), 627–634. https://doi.org/10.1002/fuce.201700026

Matko, T., Chew, J., Wenk, J., Chang, J., & Hofman, J. (2021). Computational fluid dynamics simulation of two-phase flow and dissolved oxygen in a wastewater treatment oxidation ditch. *Process Safety and Environmental Protection, 145*, 340–353. https://doi.org/10.1016/j.psep.2020.08.017

Mehl, M., Neckel, T., & Neumann, P. (2011). Navier-Stokes and Lattice-Boltzmann on octree-like grids in the Peano framework. *International Journal for Numerical Methods in Fluids, 65*(1–3), 67–86. https://doi.org/10.1002/fld.2469

Meroney, R. N., & Sheker, R. E. (2015). Removing grit during wastewater treatment: CFD analysis of HDVS performance. *Water Environment Research, 88*(5), 438–448. https://doi.org/10.2175/106143015x14212658614478

Mihalič, F., Truntič, M., & Hren, A. (2022). Hardware-in-the-loop simulations: A historical overview of engineering challenges. *Electronics, 11*(15), 2462. https://doi.org/10.3390/electronics11152462

Minin, O. (Ed.). (2011). *Computational fluid dynamics technologies and applications*. InTech. https://doi.org/10.5772/686

Morchain, J., Gabelle, J.-C., & Cockx, A. (2013). Coupling of biokinetic and population balance models to account for biological heterogeneity in bioreactors. *AIChE Journal, 59*(2), 369–379. https://doi.org/10.1002/aic.13820

Moreira, F. C., Boaventura, R. A. R., Brillas, E., & Vilar, V. J. P. (2017). Electrochemical advanced oxidation processes: A review on their application to synthetic and real wastewaters. In *Applied*

Catalysis B: Environmental (Vol. 202, pp. 217–261). Elsevier B.V. https://doi.org/10.1016/j.apcatb.2016.08.037

Moustakas, K., Rehan, M., Loizidou, M., Nizami, A. S., & Naqvi, M. (2020). Energy and resource recovery through integrated sustainable waste management. *Applied Energy, 261*, 114372. https://doi.org/10.1016/j.apenergy.2019.114372

Mrope, H. A., Chande Jande, Y. A., & Kivevele, T. T. (2021). A Review on computational fluid dynamics applications in the design and optimization of crossflow hydro turbines. *Journal of Renewable Energy, 2021*, 1–13. https://doi.org/10.1155/2021/5570848

Muhammad, N. (2021). Finite volume method for simulation of flowing fluid via OpenFOAM. *The European Physical Journal plus, 136*(10), 1010. https://doi.org/10.1140/epjp/s13360-021-01983-y

Nadal-Rey, G., McClure, D. D., Kavanagh, J. M., Cornelissen, S., Fletcher, D. F., & Gernaey, K. V. (2021). Understanding gradients in industrial bioreactors. *Biotechnology Advances, 46*, 107660. https://doi.org/10.1016/j.biotechadv.2020.107660

Naik, S., & Jujjavarappu, S. E. (2020). Simultaneous bioelectricity generation from cost-effective MFC and water treatment using various wastewater samples. *Environmental Science and Pollution Research, 27*(22), 27383–27393. https://doi.org/10.1007/s11356-019-06221-8

Nakatsuka, N., Kishita, Y., Kurafuchi, T., & Akamatsu, F. (2020). Integrating wastewater treatment and incineration plants for energy-efficient urban biomass utilization: A life cycle analysis. *Journal of Cleaner Production, 243*, 118448. https://doi.org/10.1016/J.JCLEPRO.2019.118448

Norouzi-Firouz, H., Sarrafzadeh, M., Zarghami, R., & Moshiri-Tabrizi, I. (2022). A coupled hydrodynamic-biokinetic simulation of three-phase flow in an oxidation ditch using CFD. *The Canadian Journal of Chemical Engineering, 100*(2), 223–236. https://doi.org/10.1002/cjce.24095

Oh, S. T. (2010). *Trend of mathematical models in microbial fuel cell for environmental energy refinery from waste/water* (pp. 25–30). https://doi.org/10.1007/978-3-642-13624-5_3

Ortiz-Martínez, V. M., Salar-García, M. J., de los Ríos, A. P., Hernández-Fernández, F. J., Egea, J. A., & Lozano, L. J. (2015). Developments in microbial fuel cell modeling. *Chemical Engineering Journal, 271*, 50–60. https://doi.org/10.1016/j.cej.2015.02.076

Ortiz-Martínez, V. M., Salar-García, M. J., los Ríos, A. P., Hernández-Fernández, F. J., Sánchez-Segado, S., Egea-Larrosa, J. A., & Lozano-Blanco, L. J. (2014). Avances recientes en modelado y simulación de pilas de combustible microbianas. *DYNA-Ingeniería e Industria, 89*(6).

Ou, S., Kashima, H., Aaron, D. S., Regan, J. M., & Mench, M. M. (2017). Full cell simulation and the evaluation of the buffer system on air-cathode microbial fuel cell. *Journal of Power Sources, 347*, 159–169. https://doi.org/10.1016/j.jpowsour.2017.02.031

Owusu, P. A., & Asumadu-Sarkodie, S. (2016). A review of renewable energy sources, sustainability issues and climate change mitigation. *Cogent Engineering, 3*(1), 1167990. https://doi.org/10.1080/23311916.2016.1167990

Ozgun, H., Dereli, R. K., Ersahin, M. E., Kinaci, C., Spanjers, H., & van Lier, J. B. (2013). A review of anaerobic membrane bioreactors for municipal wastewater treatment: Integration options, limitations and expectations. *Separation and Purification Technology, 118*, 89–104. https://doi.org/10.1016/J.SEPPUR.2013.06.036

Palatsi, J., Ripoll, F., Benzal, A., Pijuan, M., & Romero-Güiza, M. S. (2021). Enhancement of biological nutrient removal process with advanced process control tools in full-scale wastewater treatment plant. *Water Research, 200*, 117212. https://doi.org/10.1016/j.watres.2021.117212

Pan, H., Chen, X.-Z., Liang, X.-F., Zhu, L.-T., & Luo, Z.-H. (2016). CFD simulations of gas–liquid–solid flow in fluidized bed reactors—A review. *Powder Technology, 299*, 235–258. https://doi.org/10.1016/j.powtec.2016.05.024

Panneerselvam, R., Savithri, S., & Surender, G. D. (2007). CFD based investigations on hydrodynamics and energy dissipation due to solid motion in liquid fluidised bed. *Chemical Engineering Journal, 132*(1–3), 159–171. https://doi.org/10.1016/j.cej.2007.01.042

Partopour, B., & Dixon, A. G. (2016). Reduced microkinetics model for computational fluid dynamics (CFD) simulation of the fixed-bed partial oxidation of ethylene. *Industrial & Engineering Chemistry Research, 55*(27), 7296–7306. https://doi.org/10.1021/acs.iecr.6b00526

Patziger, M. (2021). Improving wastewater treatment plant performance by applying CFD models for design and operation: Selected case studies. *Water Science and Technology, 84*(2), 323–332. https://doi.org/10.2166/wst.2021.019

Peleka, E. N., & Matis, K. A. (2016). Hydrodynamic aspects of flotation separation. *Open Chemistry, 14*(1), 132–139. https://doi.org/10.1515/chem-2016-0014

Pereira, J. P., Karpinska, A., Gomes, P. J., Martins, A. A., Dias, M. M., Lopes, J. C. B., & Santos, R. J. (2012). Activated sludge models coupled to CFD simulations. In *Single and two-phase flows on chemical and biomedical engineering* (pp. 153–173). Bentham Science Publishers Ltd. https://doi.org/10.2174/978160805295011201010153

Pilarek, M., Sobieszuk, P., Wierzchowski, K., & Dąbkowska, K. (2018). Impact of operating parameters on values of a volumetric mass transfer coefficient in a single-use bioreactor with wave-induced agitation. *Chemical Engineering Research and Design, 136*, 1–10. https://doi.org/10.1016/j.cherd.2018.04.012

Popat, S. C., & Torres, C. I. (2016). Critical transport rates that limit the performance of microbial electrochemistry technologies. *Bioresource Technology, 215*, 265–273. https://doi.org/10.1016/j.biortech.2016.04.136

Radeef, A. Y., & Ismail, Z. Z. (2019). Polarization model of microbial fuel cell for treatment of actual potato chips processing wastewater associated with power generation. *Journal of Electroanalytical Chemistry, 836*, 176–181. https://doi.org/10.1016/j.jelechem.2019.02.001

Ratkovich, N., Horn, W., Helmus, F. P., Rosenberger, S., Naessens, W., Nopens, I., & Bentzen, T. R. (2013). Activated sludge rheology: A critical review on data collection and modelling. *Water Research, 47*(2), 463–482. https://doi.org/10.1016/j.watres.2012.11.021

Rehman, U., Amerlinck, Y., Arnaldos, M., & Nopens, I. (n.d.). *CFD and Biokinetic model integration applied to a full scale WWTP*.

Ren, J., & Cao, S.-J. (2020). Development of self-adaptive low-dimension ventilation models using OpenFOAM: Towards the application of AI based on CFD data. *Building and Environment, 171*, 106671. https://doi.org/10.1016/j.buildenv.2020.106671

Ren, P., Li, W., & Yu, K. (2021). Computational fluid dynamics simulation of adsorption process in a liquid-solids fluidized bed. *Journal of Environmental Chemical Engineering, 9*(4), 105428. https://doi.org/10.1016/j.jece.2021.105428

Rivera-Alvarez, I., Brown, R. K., Keskin-Pyttel, D., Steffens, J., Farber, P., & Schröder, U. (2020). Correlating theoretical boundary layer thickness to the power output of a microbial fuel cell with a complex anode geometry operated at varying flow rates. *Journal of Power Sources, 470*, 228428. https://doi.org/10.1016/j.jpowsour.2020.228428

Runchal, A. K., & Rao, M. M. (2020). CFD of the future: Year 2025 and beyond. In *50 years of CFD in engineering sciences* (pp. 779–795). Springer Singapore. https://doi.org/10.1007/978-981-15-2670-1_22

Saini, A. K., Paritosh, K., Singh, A. K., & Vivekanand, V. (2021). CFD approach for pumped-recirculation mixing strategy in wastewater treatment: Minimizing power consumption, enhancing resource recovery in commercial anaerobic digester. *Journal of Water Process Engineering, 40*, 101777. https://doi.org/10.1016/j.jwpe.2020.101777

Salehpour, R., Jalilnejad, E., Nalband, M., & Ghasemzadeh, K. (2020). Hydrodynamic behavior of an airlift reactor with net draft tube with different configurations: Numerical evaluation using CFD technique. *Particuology, 51*, 91–108. https://doi.org/10.1016/j.partic.2019.09.005

Samstag, R. W., Ducoste, J. J., Griborio, A., Nopens, I., Batstone, D. J., Wicks, J. D., Saunders, S., Wicklein, E. A., Kenny, G., & Laurent, J. (2016a). CFD for wastewater treatment: An overview. *Water Science and Technology, 74*(3), 549–563. https://doi.org/10.2166/wst.2016.249

Samstag, R. W., Ducoste, J. J., Griborio, A., Nopens, I., Batstone, D. J., Wicks, J. D., Saunders, S., Wicklein, E. A., Kenny, G., & Laurent, J. (2016b). CFD for wastewater treatment: An overview.

In *Water science and technology* (Vol. 74, Issue 3, pp. 549–563). IWA Publishing. https://doi.org/10.2166/wst.2016.249

Sánchez, F., Rey, H., Viedma, A., Nicolás-Pérez, F., Kaiser, A. S., & Martínez, M. (2018). CFD simulation of fluid dynamic and biokinetic processes within activated sludge reactors under intermittent aeration regime. *Water Research, 139*, 47–57. https://doi.org/10.1016/j.watres.2018.03.067

Sandhibigraha, S., Sasmal, S., Bandyopadhyay, T. K., & Bhunia, B. (2019). Computational fluid dynamics analysis of flow through immobilized catalyzed packed bed reactor for removal of 4-chlorophenol from wastewater. *Environmental Engineering Research, 25*(6), 878–889. https://doi.org/10.4491/eer.2019.184

Sangeetha, T., Li, I.-T., Lan, T.-H., Wang, C.-T., & Yan, W.-M. (2021). A fluid dynamics perspective on the flow dependent performance of honey comb microbial fuel cells. *Energy, 214*, 118928. https://doi.org/10.1016/j.energy.2020.118928

Sathe, S. M., Chakraborty, I., & Ghangrekar, M. M. (2021). Wastewater treatment and concomitant bioelectricity production using microbial fuel cell: Present aspects, up-scaling and future inventiveness. *Transactions of the Indian National Academy of Engineering, 6*(3), 633–651. https://doi.org/10.1007/s41403-021-00245-8

Satpathy, K., Rehman, U., Cools, B., Verdickt, L., Peleman, G., & Nopens, I. (2020). CFD-based process optimization of a dissolved air flotation system for drinking water production. *Water Science and Technology, 81*(8), 1668–1681. https://doi.org/10.2166/wst.2020.028

Schwedhelm, I., Zdzieblo, D., Appelt-Menzel, A., Berger, C., Schmitz, T., Schuldt, B., Franke, A., Müller, F.-J., Pless, O., Schwarz, T., Wiedemann, P., Walles, H., & Hansmann, J. (2019). Automated real-time monitoring of human pluripotent stem cell aggregation in stirred tank reactors. *Scientific Reports, 9*(1), 12297. https://doi.org/10.1038/s41598-019-48814-w

Shadloo, M. S., Oger, G., & Le Touzé, D. (2016). Smoothed particle hydrodynamics method for fluid flows, towards industrial applications: Motivations, current state, and challenges. *Computers & Fluids, 136*, 11–34. https://doi.org/10.1016/j.compfluid.2016.05.029

Shen, L., Wu, Q., Ye, Q., Lin, H., Zhang, J., Chen, C., Yue, R., Teng, J., Hong, H., & Liao, B.-Q. (2023). Superior performance of a membrane bioreactor through innovative in-situ aeration and structural optimization using computational fluid dynamics. *Water Research*. https://doi.org/10.1016/j.watres.2023.120353

Shen, R., Jiao, Z., Parker, T., Sun, Y., & Wang, Q. (2020). Recent application of computational fluid dynamics (CFD) in process safety and loss prevention: A review. *Journal of Loss Prevention in the Process Industries, 67*, 104252. https://doi.org/10.1016/j.jlp.2020.104252

Shen, Z., Zhang, M., Fan, X., Shi, S., & Han, D. (2019). Hydrodynamic and flotation kinetic analysis of a large scale mechanical agitated flotation cell with the typical impeller and the arc impeller. *Minerals, 9*(2). https://doi.org/10.3390/min9020079

Shi, Y., Wang, Z., Du, X., Gong, B., Jegatheesan, V., & Haq, I. U. (2021). Recent advances in the prediction of fouling in membrane bioreactors. *Membranes, 11*(6), 381. https://doi.org/10.3390/membranes11060381

Shirazian, S., Rezakazemi, M., Marjani, A., & Moradi, S. (2012). Hydrodynamics and mass transfer simulation of wastewater treatment in membrane reactors. *Desalination, 286*, 290–295. https://doi.org/10.1016/j.desal.2011.11.039

Singh, H., & Hutmacher, D. W. (2009). *Bioreactor studies and computational fluid dynamics* (pp. 231–249). https://doi.org/10.1007/10_2008

Singhal, A., Cloete, S., Radl, S., Quinta-Ferreira, R., & Amini, S. (2017). Heat transfer to a gas from densely packed beds of monodisperse spherical particles. *Chemical Engineering Journal, 314*, 27–37. https://doi.org/10.1016/j.cej.2016.12.124

Sobieszuk, P., Zamojska-Jaroszewicz, A., & Makowski, Ł. (2017). Influence of the operational parameters on bioelectricity generation in continuous microbial fuel cell, experimental and computational fluid dynamics modelling. *Journal of Power Sources, 371*, 178–187. https://doi.org/10.1016/j.jpowsour.2017.10.032

Song, H.-S., Cannon, W., Beliaev, A., & Konopka, A. (2014). Mathematical modeling of microbial community dynamics: A methodological review. *Processes, 2*(4), 711–752. https://doi.org/10.3390/pr2040711

Song, X., Luo, W., Hai, F. I., Price, W. E., Guo, W., Ngo, H. H., & Nghiem, L. D. (2018). Resource recovery from wastewater by anaerobic membrane bioreactors: Opportunities and challenges. *Bioresource Technology, 270*, 669–677. https://doi.org/10.1016/j.biortech.2018.09.001

Suh, J.-W., Kim, J.-W., Choi, Y.-S., Kim, J.-H., Joo, W.-G., & Lee, K.-Y. (2018). Development of numerical Eulerian-Eulerian models for simulating multiphase pumps. *Journal of Petroleum Science and Engineering, 162*, 588–601. https://doi.org/10.1016/j.petrol.2017.10.073

Sultan, T., Ahmad, Z., Hayat, K., & Chaudhry, I. A. (2022). Computational analysis of three lamp close conduit water disinfection UV reactor. *International Journal of Environmental Science and Technology, 19*(5), 4393–4406. https://doi.org/10.1007/s13762-021-03344-9

Tan, J., Ji, Y.-N., Deng, W.-S., & Su, Y.-F. (2021). Process intensification in gas/liquid/solid reaction in trickle bed reactors: A review. *Petroleum Science, 18*(4), 1203–1218. https://doi.org/10.1016/j.petsci.2021.07.007

Trifi, D., Climent, J., Arnau, R., Carratalà, P., García, M., Beltrán, I., Badenes, C., Chiva, S., & Martínez-Cuenca, R. (2023). *Design and implementation of a passive agitator to increase UV dose in WWTPs disinfection channels* (pp. 626–636). https://doi.org/10.1007/978-3-031-15928-2_55

Tu, J., Yeoh, G. H., Liu, C., & Tao, Y. (2023). *Computational fluid dynamics: A practical approach*. Elsevier.

Udoewa, V., & Kumar, V. (2012). Computational fluid dynamics. In *Applied computational fluid dynamics*.

Wang, K., Li, L., Tan, F., & Wu, D. (2018). Treatment of landfill leachate using activated sludge technology: A review. *Archaea, 2018*, 1–10. https://doi.org/10.1155/2018/1039453

Wang, Y., Brannock, M., Cox, S., & Leslie, G. (2010). CFD simulations of membrane filtration zone in a submerged hollow fibre membrane bioreactor using a porous media approach. *Journal of Membrane Science, 363*(1–2), 57–66. https://doi.org/10.1016/j.memsci.2010.07.008

Wang, Y., Chen, J., & Larachi, F. (2013). Modelling and simulation of trickle-bed reactors using computational fluid dynamics: A state-of-the-art review. *The Canadian Journal of Chemical Engineering, 91*(1), 136–180. https://doi.org/10.1002/cjce.20702

Wang, Y., Waite, T. D., & Leslie, G. L. (2013). Computational fluid dynamics (CFD) analysis of membrane reactors: Modelling of membrane bioreactors for municipal wastewater treatment. In *Handbook of Membrane Reactors* (Vol. 1, pp. 532–568). Elsevier Inc. https://doi.org/10.1533/9780857097330.3.532

Werner, S., Kaiser, S. C., Kraume, M., & Eibl, D. (2014). Computational fluid dynamics as a modern tool for engineering characterization of bioreactors. *Pharmaceutical Bioprocessing, 2*(1), 85–99.

Weyand, B., Israelowitz, M., Kramer, J., Bodmer, C., Noehre, M., Strauss, S., Schmälzlin, E., Gille, C., von Schroeder, H. P., Reimers, K., & Vogt, P. M. (2015). Three-dimensional modelling inside a differential pressure laminar flow bioreactor filled with porous media. *BioMed Research International, 2015*, 1–9. https://doi.org/10.1155/2015/320280

Wicklein, E., Batstone, D. J., Ducoste, J., Laurent, J., Griborio, A., Wicks, J., Saunders, S., Samstag, R., Potier, O., & Nopens, I. (2016). Good modelling practice in applying computational fluid dynamics for WWTP modelling. *Water Science and Technology, 73*(5), 969–982. https://doi.org/10.2166/wst.2015.565

Wu, B. (2013). Advances in the use of CFD to characterize, design and optimize bioenergy systems. *Computers and Electronics in Agriculture, 93*, 195–208. https://doi.org/10.1016/j.compag.2012.05.008

Wu, B., & Chen, S. (2008). CFD simulation of non-Newtonian fluid flow in anaerobic digesters. *Biotechnology and Bioengineering, 99*(3), 700–711. https://doi.org/10.1002/bit.21613

Wu, Q., Yan, X., Xiao, K., Guan, J., Li, T., Liang, P., & Huang, X. (2018). Optimization of membrane unit location in a full-scale membrane bioreactor using computational fluid dynamics. *Bioresource Technology, 249*, 402–409. https://doi.org/10.1016/j.biortech.2017.09.209

Xanthos, S., Gong, M., Ramalingam, K., Fillos, J., Deur, A., Beckmann, K., & McCorquodale, J. A. (2011). Performance assessment of secondary settling tanks using CFD modeling. *Water Resources Management, 25*(4), 1169–1182. https://doi.org/10.1007/s11269-010-9620-1

Xia, C., Zhang, D., Pedrycz, W., Zhu, Y., & Guo, Y. (2018). Models for microbial fuel cells: A critical review. *Journal of Power Sources, 373*, 119–131. https://doi.org/10.1016/j.jpowsour.2017.11.001

Xiao, K., Liang, S., Wang, X., Chen, C., & Huang, X. (2019). Current state and challenges of full-scale membrane bioreactor applications: A critical review. *Bioresource Technology, 271*, 473–481. https://doi.org/10.1016/j.biortech.2018.09.061

Xiao, Y., Peng, J., Sun, W., Han, H., Xie, L., & Wang, R. (2023). Study on liquid-solid fluidisation behaviour of coarse coal particles: Experiment and CFD simulation. *Chemical Engineering Research and Design, 189*, 1–13. https://doi.org/10.1016/j.cherd.2022.11.001

Xie, F., Ge, H., Liu, J., Chen, W., & Song, H. (2018). CFD and experimental studies the effect of micro-channel turbulence promoter installation on the hydrodynamic performance of submerged flat-sheet membrane bioreactor. *Chemical Engineering and Processing - Process Intensification, 127*, 28–35. https://doi.org/10.1016/j.cep.2018.03.009

Xu, C., Rangaiah, G. P., & Zhao, X. S. (2015). A computational study of the effect of lamp arrangements on the performance of ultraviolet water disinfection reactors. *Chemical Engineering Science, 122*, 299–306. https://doi.org/10.1016/j.ces.2014.09.041

Xu, H., Yang, B., Liu, Y., Li, F., Shen, C., Ma, C., Tian, Q., Song, X., & Sand, W. (2018). Recent advances in anaerobic biological processes for textile printing and dyeing wastewater treatment: A mini-review. *World Journal of Microbiology and Biotechnology, 34*(11), 165. https://doi.org/10.1007/s11274-018-2548-y

Xu, N., Fan, L., Pang, H., & Shi, H. (2010). Feasibility study and CFD-aided design for a new type oxidation ditch based on airlift circulation. *The Canadian Journal of Chemical Engineering*. https://doi.org/10.1002/cjce.20335

Xu, Q., Wan, Y., Wu, Q., Xiao, K., Yu, W., Liang, S., Zhu, Y., Hou, H., Liu, B., Hu, J., Yang, Y., & Yang, J. (2021). An efficient hydrodynamic-biokinetic model for the optimization of operational strategy applied in a full-scale oxidation ditch by CFD integrated with ASM2. *Water Research, 193*, 116888. https://doi.org/10.1016/j.watres.2021.116888

Yang, J., Wu, J., Zhou, L., & Wang, Q. (2016). Computational study of fluid flow and heat transfer in composite packed beds of spheres with low tube to particle diameter ratio. *Nuclear Engineering and Design, 300*, 85–96. https://doi.org/10.1016/j.nucengdes.2015.10.030

Yao, Y., Fringer, O. B., & Criddle, C. S. (2022). CFD-accelerated bioreactor optimization: Reducing the hydrodynamic parameter space. *Environmental Science: Water Research & Technology, 8*(3), 456–464. https://doi.org/10.1039/D1EW00666E

Ye, C.-C., Zhang, P.-J.-Y., Wan, Z.-H., Yan, R., & Sun, D.-J. (2022). Accelerating CFD simulation with high order finite difference method on curvilinear coordinates for modern GPU clusters. *Advances in Aerodynamics, 4*(1), 7. https://doi.org/10.1186/s42774-021-00098-3

Yi, Y., Xie, B., Zhao, T., Qian, Z., & Liu, H. (2020). The effect of anode hydrodynamics on the sensitivity of microbial fuel cell based biosensors and the biological mechanism. *Bioelectrochemistry, 132*, 107351. https://doi.org/10.1016/j.bioelechem.2019.107351

Yoon, S.-H. (2015). *Membrane bioreactor processes: Principles and applications*. CRC Press.

Zawawi, M. H., Saleha, A., Salwa, A., Hassan, N. H., Zahari, N. M., Ramli, M. Z., & Muda, Z. C. (2018). *A review: Fundamentals of computational fluid dynamics (CFD)*. 020252. https://doi.org/10.1063/1.5066893

Zeng, Y., Choo, Y. F., Kim, B.-H., & Wu, P. (2010). Modelling and simulation of two-chamber microbial fuel cell. *Journal of Power Sources, 195*(1), 79–89. https://doi.org/10.1016/j.jpowsour.2009.06.101

Zhan, M., You, M., Liu, L., Zhang, Y., Yuan, F., Guo, B., Cheng, G., & Xu, W. (2021). Numerical simulation of mechanical flocculation in water treatment. *Journal of Environmental Chemical Engineering, 9*(4), 105536. https://doi.org/10.1016/J.JECE.2021.105536

Zhang, M., Dong, H., & Geng, Z. (2018). Computational study of flow and heat transfer in fixed beds with cylindrical particles for low tube to particle diameter ratios. *Chemical Engineering Research and Design, 132*, 149–161. https://doi.org/10.1016/j.cherd.2018.01.006

Zhang, S., Müller, D., Arellano-Garcia, H., & Wozny, G. (2013). CFD simulation of the fluid hydrodynamics in a continuous stirred-tank reactor. In *Chemical Engineering Transactions* (Vol. 32). https://doi.org/10.3303/CET1332241

Zhao, L., Li, J., Battaglia, F., & He, Z. (2016). Investigation of multiphysics in tubular microbial fuel cells by coupled computational fluid dynamics with multi-order Butler-Volmer reactions. *Chemical Engineering Journal, 296*, 377–385. https://doi.org/10.1016/j.cej.2016.03.110

Chapter 2
CFD Modelling for Optimization of Wastewater Treatment Processes: Towards a Low-Cost Cleaner Future Tool

1 Introduction

There are several options for pollutant removal, but the preference must be based on construction costs, the source of the liquid waste, the added value of the resulting material, and the possible effects of the industry accepted on the ecosystem. Computational fluid dynamics (CFD) is a key tool for evaluating wastewater reactor configuration and nearly all unit process optimization. This chapter presents an overview of the application of CFD in wastewater treatment plants or reactors to a broader range of unit reactions in water and wastewater treatment, including bioprocess elements such as suspended growth, disinfection, and anaerobic digestion, as well as wastewater reactors optimization by using CFD simulation modelling process. The economic aspects of wastewater bioreactors have a significant impact on the development of cost-effective wastewater bioreactors. The functionality of CFD differs depending on the processing unit, with activated sludge basin modelling, final sedimentation, and disinfection seeing the most commonly used approaches, and hydrodynamics, primary sedimentation, and anaerobic digestion seeing to need further study. Existing methods provide valuable insights and can handle non-Newtonian fluids, multiphase systems, and hydrodynamics, but there are still gaps, and further development is required to accommodate the vast array of biological processes and reactor designs.

2 Background

The rapid spread of a broad range of pollutants in groundwater and surface water has become a key problem worldwide, due to the rapid development of industrialization. It is thus critical to control the deleterious effects of pollutants while also improving the human living environment (Carey & Migliaccio, 2009). Wastewater

S. E. Jujjavarapu et al., *Computational Fluid Dynamics Applications in Bio and Biomedical Processes*, https://doi.org/10.1007/978-981-99-7129-9_2

treatment plants or bioreactors are widely used across the world and are an important step in enhancing wastewater quality before it is withdrawn to the surface or groundwater and re-enters water sources. However, economic efficiency is a critical deciding factor in the choice of a technique and bioreactors between many competing technologies (Ozgun et al., 2013). The main concern of researchers in developing wastewater treatment plants technique is developing economically efficient treatment plants. Regrettably, there has not been much development to the operational process and design of bioreactors due to the high experimental time and overall capital cost of treatment plants or bioreactors. Bioreactors efficiently recycle biomass from wastewater for a variety of purposes, including energy and raw materials (Nakatsuka et al., 2020). However, process optimization in bioreactors is critical to ensuring higher energy efficiency of biomass utilization.

CFD is becoming a popular technique for evaluating flow behaviour in a variety of industries. It has gained popularity in recent years for the study of fluid problems in wastewater treatment reactors, but it still needs to gain popularity for the assessment of biological, physical, and chemical processes in wastewater treatment. CFD has key significance in terms of economics and time for better wastewater system design and biomass utilization (Wicklein et al., 2016). Biological reactors are the most energy-intensive facilities of conventional wastewater treatment plants (WWTP) due to the aeration system. Numerous biological reactors operate on an intermittent aeration regime; optimizing the aeration process is required to ensure high efficiency, meeting untreated sewage requirements with the least amount of power utilization (Sánchez et al., 2018). For improved wastewater system design for biomass utilization and waste treatment, CFD has substantial financial and risk drivers. Even so, few educational organizations are concentrating on CFD learning or research in the sewage treatment sector and guidelines are limited. Wicklein et al. (2016) are involved in the best model-based processes for wastewater CFD simulation. CFD is used to model aeration, allowing for improved treatment performance and reduced energy requirement. Several multiphase or single-phase strategies commonly used in CFD research of aeration tank process are discussed thoroughly, as they are the flaws of the model-based assumptions used to assess mixing and mass transfer in-tank reactors (Karpinska & Bridgeman, 2016b).

This chapter aims to provide a comprehensive evaluation of recent advancements in the use of CFD for analysing water and fluid properties in WWTPs. Not only does it evaluate the current state of CFD applications in this field, but it also examines critically the areas that require additional research and development. In addition, the chapter explores the techno-economic implications of implementing CFD simulation modelling techniques for WWTPs. It is crucial for decision-makers and stakeholders to evaluate the costs and benefits associated with CFD implementation in WWTPs. By taking into account factors such as computational requirements, data acquisition, model calibration, and maintenance, it is possible to conduct a thorough analysis of the economic viability and potential benefits of utilizing CFD.

3 Optimization of Bioprocess in Wastewater Treatment Plants via a CFD

Computational fluid dynamics is a rapidly growing field in sewage treatment, with applications in almost every unit operation. It has recently received much attention for analysing hydrodynamic issues in water and WWT plants (Naik & Jujjavarappu, 2020; Wicklein et al., 2016). Hydrodynamic simulation (CFD) is a key tool for organizing the renovation and operation of frameworks in WWT plants, and its widespread use is due to the gain of efficiency, and cost and time savings that can be realized (Patziger, 2021). In this section, we discussed some biological processes, such as suspended growth, disinfection, and anaerobic condition, and their modelling in the WWT plant using CFD (Table 1).

3.1 Suspended Growth (Flocculent)

Suspended growth, also known as flocculent growth, is a biological wastewater treatment process in which microbes are suspended in the wastewater to aid in the removal of organic matter and nutrients. The suspended growth process relies on the metabolism of aerobic microorganisms, which require oxygen. As the wastewater flows through the treatment system, it contacts the suspended microorganisms, allowing them to consume and degrade the organic pollutants in the water. This process reduces organic pollutants such as carbonaceous compounds and removes nutrients such as nitrogen and phosphorus. Analysing suspended growth processes is crucial in the wastewater treatment industry, and CFD is a key tool for doing so. Understanding and optimizing biological treatment systems is facilitated by the ability to investigate flow patterns, oxygen transfer, nutrient transport, floc dynamics, and reactor design.

A critical evaluation of CFD of sludge reactors was published by Karpinska and Bridgeman (2016a, 2016b). This critical evaluation examines turbulence models, multiphase simulations, and Reynolds-averaged Navier–Stokes (RANS) simulation, models. They conclude that due to the high RAM and CPU requirements, CFD simulation of complex fluid flows in sludge reactors or tanks remains a significant challenge. Despite the lack of a clear standard CFD technique, the RANS/unsteady RANS closed by a model (i.e. $k - \varepsilon$ turbulence model) strategy has been recognized as the basic activated sludge (AS) model-based strategy. Assessments of velocity and solids characteristics in AS tanks were conducted in the field of mixing simulation. It was observed that the mechanical mixing inside an AS reactor was significantly overestimated when the density couple was not included (Samstag et al., 2016b). Using 3D CFD modelling, these simulate multiphase water, air, and active solid material in an AS tank. Rehman et al. (2014) investigated the ammonium, velocity, and dissolved oxygen in a full-scale closed-loop reactor using an integrated biokinetic and hydrodynamic model. They used a CFD package to model the hydrodynamics

Table 1 Application of CFD software in bioprocess of wastewater treatment tanks

S. No.	CFD software	Modelling or simulation technique	Application	References
1	Commercial CFD package	k-ε turbulence model	To construct impeller twisting inside a multiple reference frame zone and simulate stirring with four different mixing systems	Wu and Chen (2008)
2	Commercial CFD package	k- ε turbulence model	Estimates of velocity and solids profiles of activated sludge	Samstag et al. (2016b)
3	Commercial CFD package	Eulerian two-fluid model and drift flux model	Simulate the bubble dispersion and effect of turbulence	Rehman et al. (2016)
4	Commercial CFD package	Anaerobic digestion modelling	Enhance and design more effective AS plant geometric features and mixing systems	Batstone et al. (2015)
5	Commercial CFD package	Reynolds-averaged Navier–Stokes and standard k–ε turbulence model	Increase the performance of the AD tank, and the fluid efficiency of the bioreactor	López-Jiménez et al. (2015)
6	FLUENT	Monte–Carlo method	UV disinfection of reactors with various lamp arrangements	Xu et al. (2015)
7	Commercial CFD package	Reynolds-averaged Navier–Stokes (RANS) simulations	Simulation of the complex fluid flows in sludge tanks	Karpinska and Bridgeman (2016b)
8	FLUENT	Particle minimum UV dose method and CFD	Increase the disinfection flow rate	Li et al. (2016)
9	FLUENT	Six-lamp UV disinfection arrangement and CFD	Demonstrate the flow rate in the secondary water system	Li et al. (2018a, 2018b)
10	Commercial CFD package	The population balance model (PBM) and CFD	Flocculation enhancement of activated tank	Zhan et al. (2021)

of an AS tank in three-dimensional by an Eulerian two-fluid modelling method for gas and fluid phases. To simulate bubble dispersion, a realizable drift flux model was used, and a k-ε turbulence model was used to evaluate the influence of turbulence. When the results were compared to those obtained from a full-scale bioreactor, a 15% improvement in results was demonstrated (Rehman et al., 2014).

Recently, the CFD analysis was used to calculate the retention time distribution and gradient of velocity in the local area to assess the hydrodynamic flocculation performance (Llano-Serna et al., 2019). These findings can be used to investigate the fluid flow behaviour of flocculation systems, optimize operating parameters, and architecture flocculators in an AS tank. Zhan et al. (2021) propose a coupled CFD and population balance model that was used to simulate the flocculation and flow behaviour patterns in a full-scale flocculation test. The results showed that as the inlet hydrodynamics increased, the floc particle size decreased, resulting in a decrease in flocculation performance. Finally, they conclude that the simulated results provide a complete knowledge of the flocculation development process, which is necessary for flocculation improvement (Zhan et al., 2021).

3.2 *Anaerobic Processes*

Wastewater treatment using anaerobic processes involves biological treatment methods that take place in the absence of oxygen. Microorganisms in an oxygen-depleted environment decompose and transform organic matter in these processes. Anaerobic digesters have relatively high solids content (1–3%) in a sewage treatment setting, with efficiency being highly dependent on fluid flows. The costs of the digester process, as well as its functional capabilities, are intricately bound to the mixing of the digester (Samstag et al., 2016b). Because the capital cost of a digester can be higher (because of the higher volumes of the digester), it was crucial to enhance and design more effective AS plant geometric features and mixing systems at a minimal cost. In a recent analysis, CFD was identified as an important study that needs anaerobic digestion modelling (Batstone et al., 2015). The non-Newtonian habit of solids in digesters, which is typically sheared thinning and is influenced by solid particle concentration and temperature, is a major concern in CFD. Eshtiaghi et al. (2013) conducted an in-depth investigation of this behaviour, which is typically simulated by Herschel–Bulkley or Bingham models (or a combination) and toxic waste rheology (Eshtiaghi et al., 2013). Lopez-Jimenez et al. (2015) created a 3D numerical investigation model for the anaerobic reactor sludge flow characteristics at the Ontinyent Sewage Treatment Plant. The method was founded on the Reynolds-averaged Navier–Stokes (RANS) formula, and closure was achieved using the classic standard k-ε turbulence model. The results show that flow rate has a significant impact on the performance of the AD tank, and the bioreactor's fluid efficiency was investigated using the percentage of dead areas and recirculation density as indicators (López-Jiménez et al., 2015).

Recently, the behaviour of lagoons and the accumulation of solids in AD tank was evaluated by (Ahmmed et al., 2022). It has been discovered that lagoon depth and side-wall angle have a significant impact on mixing and sludge behaviour. In terms of sludge holding capacity and minimizing internal recycles and bypass flows, a deeper lagoon with a sharp slope performs better. The study also emphasizes the significance of maintaining an appropriate hydraulic retention time (HRT) range of 100–300 days for optimal performance. According to the findings, short-term tests may be too conservative in predicting sludge accumulation rates. In comparison to the observed results, the biochemical model used in the study overpredicted sludge accumulation.

Moreover, Dabiri et al. (2023) utilized CFD as a tool for assessing WWTPs' energy performance and mixing quality. In order to incorporate the Anaerobic Digestion Model No. 1 (ADM1) into CFD, a brand-new solver named PSADM1Foam was created. In this investigation, the authors estimated methane yield and looked into the connection between mixing power usage and methane production. Increasing the recirculation flow rate increased the methane yield but also the energy required to produce the methane. However, there are significant obstacles, such as modelling multiphase flow, effectively portraying non-Newtonian fluid dynamics, accounting for complicated biological reaction kinetics, addressing scale and computing resource restrictions, and getting trustworthy validation data. Addressing these issues would increase the accuracy and reliability of CFD models, leading to better design and optimization of anaerobic wastewater treatment plants.

3.3 Disinfection

UV reactor CFD modelling is quickly becoming a standard and strategy for characterizing, building, and problem resolution in UV disinfector reactors. Furthermore, growing trust in computational modelling validated by comprehensive biodosimetry data can lead UV manufacturers to incorporate the device into an online dosage measuring algorithm. Mathematical UV disinfection approaches are complex, necessitating the proper implementation of numerous elements before the numerical data can be used for UV model development. UV disinfection of reactors with diverse lamp frameworks was performed by Xu et al. (2015) using the CFD simulation software FLUENT, for the simulation of motion of microbe particles in various UV water disinfection reactors, and the reactor efficiency was evaluated using the Monte–Carlo method which is based on microbiological log reducing at continuous UV dosage. It has been found that the overall impact on reactor log reduction is complex and difficult. However, the outcomes for various lamp frameworks demonstrate that an increase in the number of lamps did not improve reactor effectiveness (Xu et al., 2015). Li et al. (2016) introduced a novel performance indicator, particle minimum UV dose (D_{min}), that is independent of the microorganism UV dosages curve, and its viability was investigated in the design enhancement of three-lamp UV reactor designs for water disinfection and observed by both CFD simulations and theoretical deductions (Fig. 1). D_{min} values were greater in reactor designs with normal (NOR) lamp position configurations (110–193 mW cm^2) than those in reactor designs with reverse (REV) lamp position configurations (98–160 mW cm^2). The maximum disinfection flow rate (m^3 h^{-1}) of the optimally designed UV reactor (i.e. NOR-0.4-75) was found to be increased by 47–100% when compared to non-optimally designed reactors, resulting in 32–50% energy savings (Li et al., 2016).

In recent years, the use of UV disinfection in the secondary water system has become common in many countries. Regrettably, due to the extremely varying flow conditions, running UV disinfection reactors regularly is a significant waste of energy. In 2018, Li et al. developed a cost-effective approach for a six-lamp ultraviolet (UV)

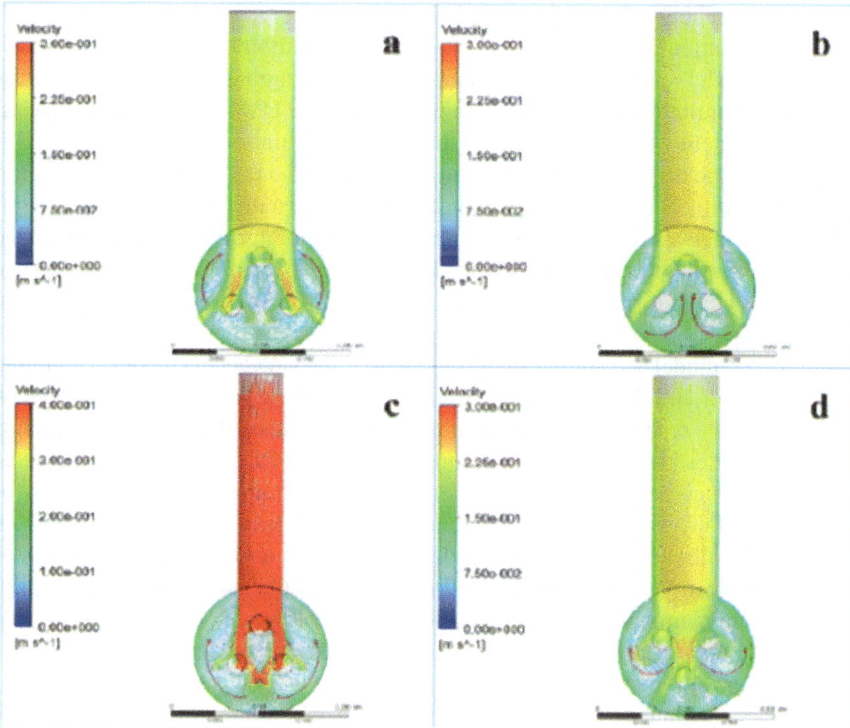

Fig. 1 CFD modelling and simulation of UV disinfection reactor or process; Flow fields at the inlet cross-sections of UV reactors with varied lamp arrangements and inlet IDs [denoted in the form of NOR (or REV)-dr-inlet ID (mm)]: **a** NOR-0.5-75, **b** NOR-0.4-75, **c** NOR-0.4-50, and **d** REV-0.4-75 (Li et al., 2016)

reactor for disinfection in a secondary water system, which consisted of multiple lamp operating conditions for varying periods, and CFD simulation was used to evaluate the reactor performance of various lamp output powers. The findings demonstrate that the rate of fluid flow in the secondary water system changed significantly during the daytime but shared common daily changing patterns over time and this created the framework for the reactor building's economical operation by lowering the lamp output power regularly. However, the defined cost-effective approach for the reactor can only save 32% of the power (Li et al., 2018a, 2018b). However, according to Trifi et al. (2023), UV radiation does not successfully reach the intermediate regions between lamps in the presence of substantial quantities of suspended particulates. Furthermore, the flow parameters in this region indicate poor mixing, resulting in restricted UV radiation exposure of pathogens. To overcome this hydraulic issue, multitubular passive agitators with turbulent vortex shedding after each tube were introduced to improve mixing. A sophisticated CFD framework was used to calculate the appropriate tube diameter. The simulation included comprehensive transient resolution of irradiation using a Monte–Carlo model, flow turbulence using a large eddy simulation model, and UV dose distribution using a Lagrangian model. Following that, the best multitubular structure was installed in the UV channel, and its impact on *E. Coli* content was assessed across multiple configurations.

Moreover, the disinfection process also depends on the positions of the UV lamps in the UV reactor. Sultan et al. (2022) investigated alternative UV lamp placements based on the vertices of four types of triangles: equilateral, right angle, scalene, and isosceles. The goal was to optimize lamp location for optimal UV dosage delivery to pathogens in water disinfection UV reactors by maximizing the reduction equivalent dosage (RED). To simulate the reactors, computational fluid dynamics was used, with the standard k-model accounting for turbulent flow, the UVCalc3D model accounting for fluence rate, and the discrete phase model accounting for pathogen movement. According to the results, scalene lamp positioning produced the highest RED value, whereas isosceles lamp positioning produced the lowest RED value. The RED value difference between equilateral and scalene light positioning was determined to be insignificant. Furthermore, the study found that using two lamps on the upstream side and one bulb on the downstream side resulted in better UV disinfection results.

CFD disinfection is a relatively well-developed application for energy savings and disinfection in wastewater treatment systems. However, no CFD model was developed to date to investigate the efficiency and performance of UV disinfection in wastewater with particle aggregates. To be able to model such a system, it may be necessary to simulate or model the continued accumulation of floc within a UV reactor, the floc particle shape, and the path length of UV light, absorbance, and potential scattering due to the presence of these aggregate particles.

4 CFD for Optimization of Wastewater Treatment Plant

A water treatment system (WTS) is an industrial method that eliminates various toxic compounds or pollutants from an incoming wastewater system using a combination of physical, physiological, chemical, and mechanical processes. WWTS treats and processes untreated sewage or contaminated water from various industries before it is withdrawn into the water system (Hreiz et al., 2019). Computational fluid dynamics is a tool that optimizes wastewater systems such as primary sedimentation, secondary sedimentation, grit removal, and dissolves air flotation to improve wastewater treatment efficiency (Fig. 2). In this section, we discussed the current state of CFD in wastewater system optimization.

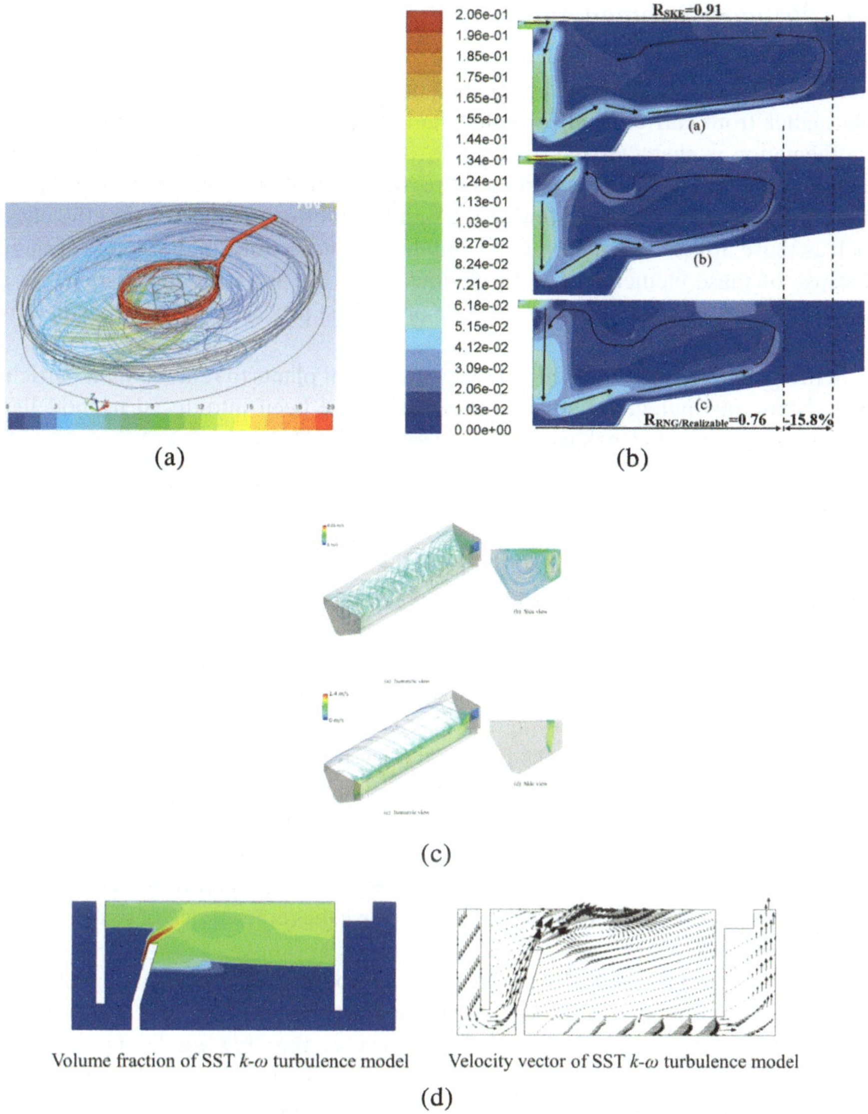

Fig. 2 Modelling and optimization of wastewater treatment plant via a CFD; **a Primary sedimentation**: flow pattern in the current clarifier illustrated by pathlines coloured by velocity magnitude (ft/min) (Das et al., 2016), **b secondary sedimentation tank:** predictions of flow patterns (m/s) for the Witney SST by the turbulence models: **a** SKE model, **b** RNG model, **c** realizable k-ε model (Gao & Stenstrom, 2018), **c grit removal**: **a, b** velocity streamlines of wastewater phase; and **c, d** velocity streamlines of air phase (wastewater flow rate = 430 L/s and airflow rate = 289 Nm^3/h) (Hoiberg & Shah, 2021) and **d dissolved air flotation**: volume fraction and velocity vector results of microbubble with SST k-ω model (Lee et al., 2020)

4.1 Primary Sedimentation

Sedimentation is a common sewage treatment technology that removes particulate matter from carrier fluid in sedimentation tanks known as clarifiers. Primary sedimentation is characterized by discrete, hindered, flocculation, and compaction settling of colloidal organic and mineral particulate matter with near-neutral buoyancy. Hydrodynamics or fluid flow has a significant effect on the system efficiency, such as those affected by multiphase solid–liquid interactions in the sludge system. Because of these elements, it is an obvious candidate for CFD application. As a result, primary sedimentation became one of the first process units to be studied by CFD investigators.

Settling tanks have been used in sewage treatment plants to remove solid particulates. Numerous numerical simulation models have been introduced to modelled settling operations and enhance tank performance. Das et al. created a new 3D unsteady CFD simulation model to increase the biomass utilization efficiency of a clarification tank that was facing underperforming and limiting holding capacity of wastewater. The CFD model was used to examine the effects of ramifications and various design changes on clarifier performance. They design three clarifiers: an inward baffle, dissipating inlets, and a submerged skirt, which resulted in more homogeneous fluid flow and biomass utilization in the clarification tank (Fig. 2a). All three designs have been shown to reduce effluent total suspended solids (TSS) by up to 80%, enhancing clarifier efficiency for biomass utilization (Das et al., 2016). To improve the TSS removal rate, small settling velocity particles should be excluded during the rebuilding of sedimentation tanks for the sewage treatment plant by employing the CFD simulation tools. In this study, the CFD simulation model was used to simulate the impact of high settling area on the TSS removal rate in three distinct layouts with constant flow rates. The simulation results found that the tank with enhanced settling area by increasing the length of the tank had the greatest TSS removal rate, with the increased settling area contributing 69% to improve the removal rate of small settling velocity particles.

The computational model for this study will be extremely helpful to researchers in designing and optimizing different commercial and industrial clarifier configurations with rotating rakes to decrease effluent TSS. However, the CFD simulating model has some limitations in that it does not account for solid particle flocculation or thermal effects on settling velocity. Future research should focus on developing thermal submodels and flocculation that can be integrated into the current model of CFD.

4.2 Secondary Sedimentation

Secondary settling tanks (SSTs) are a critical method that defines the effectiveness of the activated sludge system. However, their efficiency and performance are frequently dissatisfactory (Gao & Stenstrom, 2020). Secondary sedimentation appears to face the same challenges as primary sedimentation. Various mechanisms govern secondary sedimentation tanks, such as compressive settling behaviour, near-neutral solids density, greater underflow rates, and prominent hindered. The various features of completed effluent clarification, activated sludge thickening, sludge storage capacity, and flocculation also influence modelling objectives. Secondary sedimentation was one of the first process units investigated by CFD researchers (Fig. 2b). In 2010, John developed 2D CFD models that were calibrated in the field and the laboratory. They defined density, settling velocity, and rheology using commercial CFD software and an algebraic slip CFD model (John, n.d.). To modelling of secondary sedimentation tank, Xanthos et al. (2011) developed a 3D CFD simulated model based on the specified configuration of the established Gould II-type rectangle final settling tanks (FSTs) in Battery "E" at the New York water pollution control plants (WPCP) and calibrated and validated it with in situ data collected at the site to model secondary sedimentation tanks. The primary objective of this finding was to evaluate and predict the effects of various modelling approaches on tank parameters with greater efficiency of biomass utilization using CFD (Xanthos et al., 2011).

Recently, the CFD models were used by Gao and Stenstrom in 2020 to improve SSTs performance. In this analysis, a Fluent-based 3D numerical model is used to evaluate the potential of the SSTs and enhance their effectiveness for two large sewage treatment plants in southern California. In this study, the three distinct energy-dissipating inlet (EDI) constructions among the SSTs in these two large sewage treatment plants are analysed based on the prediction of sludge concentration profiles, and velocity distribution. In this analysis, they revealed that the Orie Albertson EDI outperforms the exclusive tangential EDI under normal flow conditions, but falls short under high flow conditions (Gao & Stenstrom, 2020).

Further work is needed to develop settling simulating models that can explain sludge behaviour across the different ranges of concentration while also accounting for turbulent fluid dynamics with measurable criteria and an improved sludge rheological model. Rather than adopting a qualitative model with parameter adjustment to improve data consistency, a repeatable and systematic measurement procedure must be characterized.

4.3 Grit Removal

Grit chambers separate large, dense particles from raw wastewater using gravity or centrifugal sedimentation. Whereas the fundamentals are very clear and simple, the multiple mechanisms and multiphase nature flow rate can make analysis difficult. Analytical grit removal assessment can be traced back to Camp's (1942) work, which developed a logical approach for calculating grit removal channels based on suitable settling behaviour (Camp, 1942). However, too much of the current approach is based on manufacturer guidelines that have little or no explanatory basis. Computational fluid dynamics is a better solution to relying on manufacturer assertion, especially when the hydrodynamic design combines centrifugal and gravity sedimentation, as in a vortex separator.

Hyre (2012) provides an in-depth analysis of three distinct grit removal tank configurations: lamella, Detritor, and forced vortex. Precise gravities of 2.65 depending on silica sand and 1.5 based on empirical evaluations from field installations, as well as a sphericity proportion of 0.65, were used to simulate grit particles of 9 distinct diameters using a continuous distribution linear model. Before injecting grit particles into the inlet, multiphase (liquid and air) simulations were conducted to stable equilibrium (Hyre, 2012). Meroney and Sheker (2015) used CFD to model the efficiency of a typical hydrodynamic vortex separator (HDVS) system in grit and sand separation. They examined the effect of total suspended solids (TSS), fluid viscosities, fluid velocity, and distribution, and particle size on the performance of the separator. Whenever the particle fall velocity (W_s) to separator inflow velocity (V_n) ratio and TSS were kept constant, CFD analysis revealed that systems of varying sizes with length scale ratios vary from 1 to 10 performed correspondingly (Meroney & Sheker, 2015).

Aerated grit tanks (AGTs) are a popular grit removal technology that creates circular fluid motion with air diffusers, allowing larger granules to gain momentum and settle down in the bottom of the container or tank. However, the efficiency of AGT is highly dependent on stimulated fluid properties and rate of flow. Hoiberg & Shah recently carried out industrial experiments on a simulated tank to quantify grit volume and particle size distribution in the tank's influent and effluent in an AGT system (Fig. 2c). The recorded data were used to validate the CFD simulated model, which was used to evaluate the impact of operational conditions and tank geometry changes on fluid flow pattern and grit capture performance. They found that moving the screw conveyor recess from its previous area to an area near the wall resulted in a 34% increase in the capture efficacy of smaller particles in the 100–150 μm size range. However, they also found that increasing the length and width of the tank did not result in a significant improvement (Hoiberg & Shah, 2021). More incremental research is expected to obtain a baffle design that improves capture efficiency. Additional research could look into the organic particles found in wastewater.

Numerous desirable components of great CFD analysis have been included in current research on grit removal, including adjustment or validation activities, 3D approach, and multiphase assessment. The most significant work appears to have included solids transport evaluation and discrete settling. However, it is unclear whether or not underflow concentration levels in the AGT system have been observed. For a long time, it was supposed that grit settlement was not density-dependent.

4.4 Dissolved Air Flotation

Dissolved air flotation (DAF) has recently received much attention as a separation technique in both drinking and industrial wastewater treatment. However, the procedure, and the preceding flocculation phase, are complicated and poorly understood. So, because the procedure is multiphase, hydrodynamic studies are critical for understanding and optimizing the DAF in terms of design and operation. Amato and Wicks (2009) reported the findings of a CFD analysis of DAF for particle segregation from a precoagulated surface water system. Using multiphase Euler–Euler and k-turbulence modelling, commercial CFD software (ANSYS Fluent v.6.3.26) has been used to simulate biomass, fluid, and air hydrodynamics. The CFD model was used to obtain the volume-weighted average vorticity as a determination of floc-bubble agglomeration disturbance, with an average floc-bubble particle size of 148 m in the clarifying boundary. Floc properties were deduced from bubble accumulation, so these assertions have been used to determine the position of the "white water level", which previous research had linked to an increased solids removal rate (Amato & Wicks, 2009).

Previous research tended to focus on fluid dynamic behaviours to enhance the efficiency of the purification system of DAF. Nonetheless, the experimental application on a small scale revealed a few technological gaps in the DAF. To compensate for technological shortcomings, a simulation model based on CFD was performed to enhance the understanding of internal fluid flow and velocity properties in the DAF system. However, in most research, the standard model (i.e. k-ε turbulence model) was used without any proper consideration procedure. So, Lee et al. (2020) investigated the significant impacts on internal flow behaviours for an effective numerical model of DAF whenever a different turbulence and microbubble variables simulating model was used (Fig. 2d). They found that when the CFD approach is utilized properly during the construction procedure of a DAF system, the suitable values for the inlet microbubble variables for efficient pollutant removal efficiency in wastewater can be defined (Lee et al., 2020). Satpathy et al. recently investigated a thorough computational evaluation for the advancement and optimization of a wastewater treatment plant in Kluizen, Belgium. CFD was used to characterize multiphase flow using the three-dimensional numerical model with a scaled-down model for the design of the DAF system. They found that the properties of flow increase confidence in DAF simulation modelling, but flocs escape through the outlet from the system, which decreases the performance efficiency and reliability. To enhance the quality and ease of operation of the DAF system, modifications in system configuration, involving the use of a perforated tube for extraction of water have been proposed, and the result was found to be beneficial (Satpathy et al., 2020). However, bubble–particle collisions, the influence of particle relaxation time on bubble–particle coalescences, and other issues remain unresolved. Furthermore, additional measurement initiatives with data processing will be a suitable choice for some other fields of research.

5 CFD Modelling and Design of Wastewater Treatment Reactor

CFD is a beneficial tool for planning the renovation and operation of systems in wastewater treatment plants, and it is widely used because of the savings and performance improvements that can be recognized. This section primarily focused on flow simulation, which greatly aids in the operation and configuration or design of structures through simple means that have better biomass utilization and wastewater treatment efficiency.

5.1 Large-Scale Flotation Reactor

Flotation technology is broadly used in the treatment of contaminated water or wastewater. A flotation tank is filled with solid particle-contaminated water, which is then blended with small-diameter gas bubbles. Flow mixing improves bubble distribution and retains solid particles suspended, although this suspension maybe not have been uniform, with larger-size particles striving to remain inside the lower sections of the flotation tank. The flotation tank is divided into two main zones: the first zone is the "quiet" flotation zone and the second zone is the "noisy" reaction zone. Released bubbles come into direct contact with incoming polluted water in the reaction zone and particles and bubbles cling to each other. The flotation zone allows the particle or bubble aggregates to increase and separate the solids from the main water system (Peleka & Matis, 2016).

A difficult key concern for most flotation tanks would be how the hydrodynamics affect the particle/bubble and biomass utilization inbound water tank system. Kostoglou et al. (2007) used CFD modelling to assume the behaviour of flotation tanks used in conventional wastewater treatment plants. They used a tank with no external means of flow mixing, such as a DAF water tank, and these flotation processes are analysed, as well as the variability of their relative importance in relation to fundamental flotation process conditions. To allow faster computation, feasible assumptions have been made about the model's hydrodynamics and particle conservation issues with CFD simulations have been operating in a 2D frame of reference. The modelling revealed the viability of a complex relationship among the tank's local flotation rates and hydrodynamics, resulting in particle removal rate efficiency that flotation assumptions alone could never achieve. As a result, combining flotation theories for local flotation rates with computational fluid dynamics simulation appears to be a suitable tool in the development and design of large-scale DAF devices (Kostoglou et al., 2007). Another modelling method, Eulerian two-fluid was used to characterize a large-scale flotation cell constructed by the Beijing General Research Institute of Mining and Metallurgy (BGRIMM). CFD modelling was used to analyse the influence of the speed of the impeller and the flow of gas in the flow fields of the large-scale flotation cell. They found that this modelling method is appropriate for large-scale flotation cells and that the model could provide valuable information for research into the operation and design of flotation cells.

In a study, Kwon et al. (2006) utilized CFD models in conjunction with Acoustic Doppler Velocimetry techniques to investigate the effect of aspect ratio on flow characteristics and bubble distribution in a full-scale dissolved air flotation system. Similarly, Behin and Bahrami (2012) simulated and analysed the fluid flow regime of a dissolved air flotation vessel using a CFD model. In a comprehensive review, Basavarajappa and Miskovic (2016) summarized important studies and proposed a refined QMOM model using a corrected CFD-PBM model to investigate gas characteristics in a flotation cell. However, while these studies focused predominantly on the hydrodynamic aspects of flotation tanks, relatively little research has been conducted on oil removal efficiency. He et al. (2013) utilized CFD software to investigate the volume fraction distribution of oil, gas, and water in a flotation tank, and then derived oil removal efficiency based on the simulation results. However, the study did not assess the effect of the interaction between bubbles and oil droplets on the oil removal efficacy.

Recently, the hydrodynamics and flotation kinetics assessments for the latest and biggest flotation system in China, with a system volume of 680 m^3, were performed using CFD simulation. Based on flow pattern, gas distribution, and solid suspended evaluation, it is illustrated that a flotation cell system that contains a typical impeller generates appropriate hydrodynamics for the particle of minerals. The new advanced arc impeller has a greater collision probability than a standard impeller, resulting in improved flotation effectiveness (Shen et al., 2019). This powerful CFD software can rapidly predict the performance of new designs through hydrodynamic optimization and process optimization. On the other hand, the model in CFD is also an important factor that influences the working efficiency of floating reactors. In an experiment, Huang and Long (2020) investigated the parameters that influence the oil removal effectiveness of large-scale flotation tanks. It discusses the creation of a modified CFD model (i.e. the Bloom–Heindel model) that influence the separation performance of floating tank for oil separation. The collision and attachment efficiencies between oil droplets and bubbles in a flotation tank are included in the model. The article also includes experimental results and numerical simulations that investigate the impacts of gas flow rate, oil diameter, and oil concentration on oil removal efficiency. The redesigned CFD model is assessed and compared to an existing model using experimental findings. The study concludes that the new model can properly forecast the flotation tank's separation performance.

5.2 *Membrane Reactors*

The membrane bioreactor (MBR) technique can be employed for membrane iteration of AS for treating polluted wastewater, producing high effluent efficiency while leaving a small environmental footprint. However, membrane fouling is a major deterrent to using MBRs in wastewater treatment tanks, and it is difficult to obtain accurate velocity profile values inside the reactor in real-time due to the compact architecture of MBR membranes, which complicates MBR design/optimization. CFD analysis allows for the quantitative measurement of fluid flow dispersion values in MBRs that are not visible in experimental studies. The optimization of MBRs necessitates an understanding of membrane fouling, biokinetics, and mixing. However, most research has focused on fouling and biokinetics. Current design methods for achieving a proper flow within MBRs are largely based on assumptions and empirical techniques. However, it is hard to estimate how sludge rheological properties and vessel layout affect fluid properties, and thus overall performance, in full-scale installations. CFD is a technique for predicting how vessel characteristics and mixing electricity consumption affect fluid properties (Brannock et al., 2010). Itoh and Mimura (2013) proposed a CFD model that accounts for the temperature, velocity distributions, and concentration caused by flow properties, heat, and mass transfer in the palladium membrane reactor (PMR). A PMR has been used to recover hydrogen from cyclohexane as a promising chemical hydrogen carrier. According to the CFD analysis, substantial concentration and temperature distributions form in both the axial and radial orientations. Besides that, it is revealed that the advanced multi-tube simulating model can be used to change the reactor configuration, such as the length of the catalyst-packed layer, membrane aspect, and the operation factors such as feed rate, pressure, and temperature (Itoh & Mimura, 2013). A comprehensive MBR CFD model would predict the effect of chemical and biochemical transformation kinetic model, membrane filtration characteristics, fouling, and fluid dynamic regimes on MBR performance, as well as the plant's energy consumption as a function of operating and design choices (Wang et al., 2013a, 2013b).

Recently, Jin et al. (2019) used CFD models in combination with cold model PIV experimental investigations to optimize the efficiency and parameter of membrane bioreactors. The influences of membrane module height, aeration tube quantity, and membrane spacing on liquid phase volume flow rate, air hold-up, and shear at the surface of the membrane have been studied. An ideal architecture was defined based on an MBR with 40 mm spacing among both membranes and seven aerating tubes to introduce gas into the system. Because of this configuration, the greatest shearing force was produced at the base of the membrane module, 250 mm from the aeration tube (Jin et al., 2019). The established computational fluid dynamics model will be appropriate to attaining intrinsic phenomena and quantitatively evaluating reactor performance increases, and will thus be a very useful tool for evaluating novel concepts, scaling up, and optimizing the membrane reactor configuration.

5.3 Activated Sludge Channel Reactor

One of the critical problems in the sewage treatment industry is the productive configuration and reliable performance of the treatment process that ensures high treatment efficiency to meet effluent quality standards while retaining investment and operating costs to a minimum (Karpinska & Bridgeman, 2016b). Activated Sludge

Models (ASM) are commonly used in reactor architecture to define the biological methods in an Activated Sludge system. The ASMs were developed by Pereira et al. (2012) to better understand the activated sludge processes as well as for architecture purposes. The ASM is commonly accomplished without taking into consideration the exact macro-mixing of the activated sludge process, defeating the reason for using ASMs for design. This study investigated the production of macro-mixing data for AS processes and developed a method for combining macro-mixing with the ASM. Many activated sludge reactor designs operate on an intermittent aeration regime; optimizing the aeration process variables such as aeration length, air diffuser design, and rate of airflow per diffuser is supposed to ensure effective operation (Pereira et al., 2012). On the other hand, Sánchez et al. (2018) created a CFD simulation model of an activated sludge reactor that operates within an intermittent aeration regime. CFD modelling is used to choose an ASRs aeration structure, and two distinct aeration structures are designed to simulate. This study assessed the aeration power expenditure required to meet untreated sewage requirements and discovered a 2.8% improvement in energy consumption by reconfiguring the air diffuser design (Sánchez et al., 2018).

Hreiz et al. (2019) evaluate the effect of bubble aerator designs on the fluid dynamics of activated sludge reactors using CFD simulation models. Considering the greater gas flow rates used during AS processes, the outcomes showed that the fluid flow behaviour is highly sensitive to aerator configuration. The design with diffusers installed across the reactor basement produced the minimum dispersive flow, i.e. features similar to those of a perfect plug-flow reactor. This fluid flow had the least average turbulent dispersion and also the most uniform axial velocity and turbulence areas. Because it decreases contaminant dilution via axial dispersion and restricts raw wastewater channelling to the exit, such a fluid flow is designed to be significantly profitable for treatment processes (Hreiz et al., 2019). This article frequently employs CFD analysis and arithmetical RTD (residence time distribution) experiments to define the impacts of bubble aerator design on the hydrodynamics in activated sludge reactors.

CFD is used for modelling of aeration, allowing for improved treatment productivity and low energy input. However, several kinetics parameters, such as biological growth and the effect of fluid flow in biological systems, must be addressed in the future to improve the efficiency of waste treatment and biomass utilization in activated sludge reactors.

5.4 Continuous Solar-Collector-Reactors

The two main challenges that will need to be resolved in the future decades are the energy crisis and the scarcity of water. One of the most notable recent developments is the solar detoxification and mineralization of harmful industrial wastes to recycle and reuse various wastewater systems (Malato et al., 2009). A continuous solar-collector-reactor is used for sewage clarification via the photo-Fenton process mechanism. The continuous solar-collector-reactor is made up of parallel interconnected channels that are open on the upper side to receive sunlight. The solar photo-Fenton method has been widely used in the treatment of various organic contaminants, including a wide range of dyes, pharmaceuticals, and other pollutants (Doumic et al., 2015; Moreira et al., 2017). Dutta et al. (2019) proposed the design for modelling of hydrodynamic and efficiency characterization of economically solar-collector-reactor. The hydrodynamic characteristics of the reactor were revealed using RTD assessment and CFD simulation. Finally, the reactor's performance was evaluated in the treatment of simulated wastewater using two models for two micropollutants: Dichlorvos pesticide and Trypan Blue dye. According to CFD, dead zones are mainly focused on the two sides of the channel splitting plates and the hydrodynamics. The contaminants were found to degrade by more than 94% under ideal parameter situations, demonstrating the effectiveness of the suggested configuration (Dutta et al., 2019). In addition to the simulation model for wastewater, the reactor system was evaluated in the treatment of sewage from a jute dyeing unit, achieving a COD reduction of 58%. The solar collector reactor is a well-known technique for using natural resources to treat wastewater on a low-cost basis. Furthermore, the CFD technique was found to be advantageous in terms of economic basis and time-saving modelling method for the simulation of the solar-collector-reactor.

6 Hydrodynamics and Mass Transfer Simulation for Other Wastewater Reactors

The CFD modelling process is based on fluid dynamic principles and utilizes numerical models and algorithms to fix fluid flow problems. Hydrodynamics, or fluid flow, is important in all wastewater systems because it has a direct effect on microbial growth and population in the wastewater tank (Kumar et al., 2021). For this reason, assessing the fluid properties of a wastewater system tank or reactor is becoming increasingly popular to improve the efficiency of waste management in a water system.

A major focus of the research in wastewater treatment in water sources has been the process to develop effective extraction techniques for toxic compound separation from polluted water. Shirazian et al. (2012) reported for the first time that ammonia was separated from a liquid solution in hollow-fibre membrane reactors (HFMRs) using CFD simulation. The primary benefit of HFMRs over traditional separation methods is that they would provide a dispersion-free interaction of two phases (Shirazian et al., 2012). Furthermore, the flow velocity of the phases in the reactor can differ independently, with no flooding or unloading issues (Chang et al., 2013). Moullec et al. (2010) studied hydrodynamics, mass flow, and biological reactions in a sewage gas–liquid cross-flow reactor using CFD simulation. Ammonium, nitrate, chemical oxygen demand, and oxygen concentration levels were evaluated and compared to simulation model characteristics along the length of the reactor. This model should contribute for hydrodynamics, oxygen transfer from gas to liquid phase, transport from liquid to flocs, and physiological responses. The microscale hydrodynamic is critical for optimum process research because they have a significant impact on the reactor's overall performance (le Moullec et al., 2010). The flow characters in the continuous stirred-tank reactors of the co-operative Research Center InPROMPT, coordinated by the Technische Universitat Berlin, are depicted using 3D CFD simulation models. The findings demonstrate the presence of a high flow velocity toroidal area in various horizontal parts of the continuous stirred-tank reactors without baffles, in contrast to the widely held belief that the flow velocity magnitude in a continuous stirred-tank reactor at the final stable state is completely uniform. Furthermore, at a rotation speed of 400 rpm, the baffles inside the CSTR can substantially reduce the magnitude of the average tangential velocity, inhibiting the development of a surface vortex (Zhang et al., 2013). Dehnavi et al. have made significant progress in another study of hydrodynamic modelling of gas-fluidized reactor. This author used a two-fluid Eulerian CFD model with closure connections based on granular fluid kinetic theory. According to the simulation results, the two-fluid Eulerian model is appropriate for modelling hydrodynamics of fluidized bed reactors (Dehnavi et al., 2010). The hydrodynamic behaviour of four layouts of an airlift reactor (AR) with a net draft tube (NDT) of various net meshes (AR-NDT-3, 6, 12, and AR) has been designed and simulated for a range of airflow rates in terms of air holdup, fluid velocity, and oxygen mass transfer coefficient. For CFD analysis of fluid dynamics with Eulerian descriptions for the fluid and gas phases, the two-fluid model formulation coupled with the k–ε turbulence model was used. When the performance of the ALR-NDTs was compared to that of conventional ALR, the hydrodynamic parameter values were found to be improved with the addition of the central NDT (Salehpour et al., 2020). Currently, Luke et al. used computational fluid dynamics to investigate the topography of systematic packing and its consequences on fluid hydrodynamic and mass transfer characteristics. The computational fluid dynamics predictions for fluid dynamics and mass transfer were found to be very close to the laboratory experiment holdup data, with a 6 and 8% variance, respectively. The findings of this study emphasize the significance of capturing surface revival at organized packing crimps and at the point of contact among organized packing layers (Macfarlan et al., 2021).

Engineers and researchers can gain insight into the complex fluid dynamics and transport phenomena occurring within various wastewater reactors by employing hydrodynamics and mass transfer simulations. This knowledge allows them to optimize reactor design, enhance treatment efficiency, and develop sustainable and cost-effective wastewater treatment solutions. Hydrodynamics in the MFC has a significant impact on substrate dispersion and mass transfer rates (Kumar et al., 2021). The flow rate or hydrodynamics has a direct impact on the mass transport of microbes towards the reactor surface, which can influence microorganism physiology and behaviour, ultimately changing the efficiency of reactor performance for biomass utilization and wastewater treatment.

7 Techno-Economic Analysis

In green technology, WWTPs play an important role in increasing the efficiency of biomass utilization, cleaning the ecosystem, and the surrounding environment, particularly water system. From a techno-economic perspective, CFD plays an important role in developing a low-cost reactor without the prior development of a reactor. The market value of simple MBRs was estimated to be around US$217 million, with a per-year growth rate between about 9.5 and 12% (Wang et al., 2013a, 2013b). Depending on the economic comparing method of analysis discussed by Arif et al (2020), the results revealed that in the specified number of criteria and over 39 years, the total cost of the project of the CAS plant is $27,900,000, and the MBR plant ($63,600,000) (Arif et al., 2020).

The AS treatment method is a type of suspended growth biological treatment process used to treat polluted water systems. Economic modelling and estimation of the cost of AS processes are critical for wastewater treatment plant construction, design, and forecasting future economic requirements. However, the capital and maintenance costs of a wastewater treatment plant or reactor are much higher, making their widespread use in all contaminated areas difficult. Jafarinejad studied three AS systems in 2016 that included extended aeration activated sludge (EAAS), conventional activated sludge (CAS), and sequencing batch reactor (SBR) to estimate the total project construction, operation labour, maintenance, material, chemical, power, and amortization costs. The estimated cost for operation and maintenance was $305,500/year (CAS), $323,100 (SBR), $ 289,400 (EAAS). The results showed that increasing the mixed liquor suspended solid reduces the total project construction of WWTPs with EAAS and CAS (Jafarinejad, 2017).

However, the computational approach such as CFD reduces the estimated cost of construction, design, and future economic requirements of the WWTPs. CFD modelling is a low-cost optimization technique. In summary, computational modelling is used in bioreactor applications to optimize the operational parameters of various reactors as well as the reactor configuration. Typically, reactor research relies on a variety of laboratory experiments as well as expensive chemicals and equipment. These laboratory experiments run for a longer time. As a result, CFD modelling will provide a simple, quick, and low-cost method for optimizing process parameters that are challenging to do in real-time.

CFD software significantly reduces the cost of designing wastewater treatment plants by providing valuable insights and optimizing the design procedure. CFD software contributes to cost reduction in the following ways:

A. CFD simulations permit engineers to evaluate the performance of various design configurations and operational parameters without the need for physical prototypes. Early in the design phase, the software aids in identifying potential problems and optimizing the system's performance by virtually testing various design alternatives. This reduces the cost of physical testing and the danger of costly design changes in the future.

B. Using CFD software, engineers can optimize processes within wastewater treatment facilities, including flow distribution, mixing, and oxygen transfer. By simulating and analysing these processes, the software assists in spotting areas of inefficiency or poor performance. Optimizing these factors ensures that the treatment plant operates at peak efficiency, thereby reducing energy consumption and operating expenses.
C. Equipment design: CFD software aids in the design and optimization of equipment utilized in wastewater treatment facilities, such as pumps, mixers, and aeration systems. The software can evaluate fluid flow patterns, pressure distributions, and thermal transfer within these components, ensuring their optimal design and functionality. By selecting the proper equipment and optimizing its operation, costs associated with maintenance, energy consumption, and equipment failure can be diminished.
D. Troubleshooting and risk mitigation: In the event of operational problems or inefficiencies in an existing wastewater treatment plant, CFD simulations can aid in troubleshooting and determining the fundamental causes. By analysing fluid flow, mingling, and other factors, the software assists in identifying problem areas and recommending possible solutions. This decreases the cost of trial-and-error repairs and enhances the overall performance of the plant.
E. Design validation and compliance: CFD software aids in validating the design of wastewater treatment plants to ensure regulatory compliance. By simulating and analysing various parameters, such as effluent quality, pollutant dispersion, and odour control, the software helps affirm that the design complies with the necessary environmental standards. The costs of fines, penalties, and retrofitting can be avoided by avoiding noncompliance issues.
F. Visualization and communication: CFD software provides visual representations of fluid flow and other physical phenomena in a wastewater treatment facility. These visualizations aid in conveying design concepts, operational strategies, and performance predictions to stakeholders, including clients, regulatory bodies, and project teams. Improved communication results in enhanced decision-making, decreased error rates, and streamlined project implementation.

Utilizing the capabilities of CFD software, engineers can optimize the design and operation of wastewater treatment facilities, resulting in enhanced performance, decreased energy consumption, and reduced operating costs. The software's capability to simulate complex fluid dynamics and optimize processes enables cost-effective design decisions and the effective operation of treatment facilities.

8 Future Prospective

Using a low-cost simulation method, CFD was discovered to be extremely useful for modelling the reactor with little prior knowledge or experimentation. The above research reveals that the evaluation of sludge blankets for primary and secondary settling tanks has received little attention (Das et al., 2016). Future studies should explore a 3D CFD model that uses the whole tank domain and also includes discrete and compressive settling as well as enhanced flocculation and rheology modelling techniques (Eshtiaghi et al., 2013; John, n.d.). CFD combined with activated sludge physiochemical modelling techniques has been shown to be far superior to simplified mixed tank designs. Because solids settlement influences the density of the solid particles, and velocity characteristics, as well as the resulting catalyst concentration levels, density-coupled solid particles transport, should be incorporated in future suspended growth reactor CFD models (Jafarinejad, 2017). The use of CFD models for the verification of unique biological models based on relatively simple hydrodynamic assumptions will remain necessary (Chang et al., 2013; le Moullec et al., 2010).

Several key areas of development and application for CFD software in wastewater treatment facilities portend a bright future. Here are a few possible future directions for CFD software in this industry (Runchal & Rao, 2020):

A. Advanced process modelling: CFD software will continue to evolve, allowing for more precise and comprehensive modelling of complex processes in wastewater treatment facilities. This includes the incorporation of multiphase flow, biofilm growth, chemical reactions, and sophisticated particle tracking techniques. These improved process models will provide a greater comprehension of the treatment processes, allowing for improved design and optimization (Kumar & Jujjavarapu, 2023a, 2023b; Palatsi et al., 2021).
B. The integration of AI techniques with CFD: CFD software can facilitate more efficient and automated design optimization. Large quantities of CFD simulation data can be analysed using machine learning algorithms to identify patterns and optimize process parameters. Models utilizing artificial intelligence can aid in predicting system performance, optimizing energy consumption, and improving overall treatment efficacy (Ren & Cao, 2020).
C. Real-time monitoring and control: In wastewater treatment facilities, CFD software coupled with sensor data and real-time monitoring systems can facilitate real-time process control. The software can provide automated control systems with feedback by continuously analysing flow patterns, pollutant distributions, and other parameters. This real-time control can optimize treatment processes, boost energy efficiency, and improve system dependability (Guzmán et al., 2018; Schwedhelm et al., 2019).
D. The integration of CFD software with virtual reality (VR) and augmented reality (AR) technologies: This can provide immersive and interactive experiences for operators and engineers. This integration enables users to visualize and manipulate complex flow patterns, equipment, and process data. Virtual reality and augmented reality can aid in the training of operators, the virtual exploration of treatment plant designs, and the facilitation of efficient maintenance and troubleshooting (Lin et al., 2019).
E. Sustainability and resource recovery: Future CFD software advancements will likely concentrate on optimizing sustainable practices and resource recovery in

wastewater treatment facilities. This includes modelling energy-efficient aeration systems, nutrient recovery processes, and anaerobic digestion optimization for biogas production. Simulations of CFD can aid in the design of environmentally favourable treatment processes and the identification of resource recovery opportunities (Moustakas et al., 2020; Saini et al., 2021).

F. The integration of CFD software with Internet of Things (IoT) devices and sensors can improve data collection, monitoring, and control capabilities. Incorporating sensor data in real-time into CFD simulations enables dynamic modelling and optimization. This integration can facilitate predictive maintenance, energy management, and remote monitoring of the performance of a treatment plant (Karn et al., 2023).

9 Conclusion

As the accessibility and availability of commercial and open-source software suites have increased in recent years, the use of CFD has developed into a robust and accurate technique for designing, optimizing, and controlling wastewater treatment plant systems. An investigation of the overall application of CFD in wastewater treatment plants disclosed that the method's sophistication varies greatly between components. CFD simulation of secondary sedimentation in general and especially is well established, whereas research in anaerobic digestion is less developed. The advancement of rector for treating wastewater and increasing the efficiency of biomass utilization is currently a major issue due to high operational and maintenance costs. However, CFD was discovered to be extremely useful for modelling the reactor with little prior knowledge or experiment using a low-cost simulation method. The hydrodynamic behaviour of the reactor is an important parameter to model to improve the biological reaction and allocation of biomass present in wastewater. This clearly shows that there are possibilities to expand the significance of CFD in the wastewater evaluation process and increase the efficiency of biomass utilization.

References

Abali, B. E. (2019). An accurate finite element method for the numerical solution of isothermal and incompressible flow of viscous fluid. *Fluids, 4*(1), 5. https://doi.org/10.3390/fluids4010005

Afshari, F., Zavaragh, H. G., Sahin, B., Grifoni, R. C., Corvaro, F., Marchetti, B., & Polonara, F. (2018). On numerical methods; optimization of CFD solution to evaluate fluid flow around a sample object at low Re numbers. *Mathematics and Computers in Simulation, 152*, 51–68. https://doi.org/10.1016/j.matcom.2018.04.004

Ahmad, T., & Zhang, D. (2020). A critical review of comparative global historical energy consumption and future demand: The story told so far. *Energy Reports, 6*, 1973–1991. https://doi.org/10.1016/j.egyr.2020.07.020

Ahmmed, M. S., Skerman, A., & Batstone, D. J. (2022). Predicting long-term solid accumulation in waste stabilisation lagoons through a combined CFD-process model approach. *Chemical Engineering Research and Design, 184*, 267–276. https://doi.org/10.1016/j.cherd.2022.06.012

Alidrisi, H., & Demirbas, A. (2016). Enhanced electricity generation using biomass materials. *Energy Sources, Part A: Recovery, Utilization, and Environmental Effects, 38*(10), 1419–1427. https://doi.org/10.1080/15567036.2014.948647

Alobaid, F., Almohammed, N., Massoudi Farid, M., May, J., Rößger, P., Richter, A., & Epple, B. (2022). Progress in CFD simulations of fluidized beds for chemical and energy process engineering. *Progress in Energy and Combustion Science, 91*, 100930. https://doi.org/10.1016/j.pecs.2021.100930

Aly, H. S., Eldrainy, Y. A., Saqr, K. M., Lazim, T. M., & Jaafar, M. N. M. (2010). A mathematical model for predicting spray atomization characteristics in an Eulerian-Eulerian framework. *International Communications in Heat and Mass Transfer, 37*(6), 618–623. https://doi.org/10.1016/j.icheatmasstransfer.2010.02.003

Amato, T., & Wicks, J. (2009). *Dissolved air flotation and potential clarified water quality based on computational fluid dynamics modelling CFD with reactions-simulation of wastewater treatment plants view project*. https://doi.org/10.13140/2.1.2605.2480

Arif, A. U. A., Sorour, M. T., & Aly, S. A. (2020). Cost analysis of activated sludge and membrane bioreactor WWTPs using CapdetWorks simulation program: Case study of Tikrit WWTP (middle Iraq). *Alexandria Engineering Journal, 59*(6), 4659–4667. https://doi.org/10.1016/j.aej.2020.08.023

Arumuga Perumal, D., & Dass, A. K. (2015). A review on the development of lattice Boltzmann computation of macro fluid flows and heat transfer. *Alexandria Engineering Journal, 54*(4), 955–971. https://doi.org/10.1016/j.aej.2015.07.015

Azarpour, A., Rezaei, N., & Zendehboudi, S. (2021). Performance analysis and modeling of catalytic trickle-bed reactors: A comprehensive review. *Journal of Industrial and Engineering Chemistry, 103*, 1–41. https://doi.org/10.1016/j.jiec.2021.04.020

Bach, C., Yang, J., Larsson, H., Stocks, S. M., Gernaey, K. V., Albaek, M. O., & Krühne, U. (2017). Evaluation of mixing and mass transfer in a stirred pilot scale bioreactor utilizing CFD. *Chemical Engineering Science, 171*, 19–26. https://doi.org/10.1016/j.ces.2017.05.001

Balakin, B. V., Hoffmann, A. C., Kosinski, P., & Rhyne, L. D. (2010). Eulerian-Eulerian CFD model for the sedimentation of spherical particles in suspension with high particle concentrations. *Engineering Applications of Computational Fluid Mechanics, 4*(1), 116–126. https://doi.org/10.1080/19942060.2010.11015303

Basavarajappa, M., & Miskovic, S. (2016). Investigation of gas dispersion characteristics in stirred tank and flotation cell using a corrected CFD-PBM quadrature-based moment method approach. *Minerals Engineering, 95*, 161–184. https://doi.org/10.1016/j.mineng.2016.06.026

Batstone, D. J., Puyol, D., Flores-Alsina, X., & Rodríguez, J. (2015). Mathematical modelling of anaerobic digestion processes: Applications and future needs. *Reviews in Environmental Science and Biotechnology, 14*(4), 595–613. https://doi.org/10.1007/s11157-015-9376-4

Behin, J., & Bahrami, S. (2012). Modeling an industrial dissolved air flotation tank used for separating oil from wastewater. *Chemical Engineering and Processing: Process Intensification, 59*, 1–8. https://doi.org/10.1016/j.cep.2012.05.004

Bhanduvula, S. (2012). Finite difference method in computational fluid dynamics. *Inteternational Journal of Education and Applied Research, 2*.

Bhaskaran, R., & Collins, L. (2020). *Introduction to CFD basics*.

Bhatti, M. M., Marin, M., Zeeshan, A., & Abdelsalam, S. I. (2020). Editorial: Recent trends in computational fluid dynamics. *Frontiers in Physics*. https://doi.org/10.3389/fphy.2020.593111

Bhuyar, L. B., Thakre, S. B., & Ingole, N. W. (2009). Design characteristics of curved blade aerator w.r.t. aeration efficiency and overall oxygen transfer coefficient and comparison with CFD modeling. *International Journal of Engineering, Science and Technology* 1(1). www.ijest-ng.com

Boas, J. V., Oliveira, V. B., Simões, M., & Pinto, A. M. F. R. (2022). Review on microbial fuel cells applications, developments and costs. *Journal of Environmental Management, 307*, 114525. https://doi.org/10.1016/j.jenvman.2022.114525

Bouras, H., Haroun, Y., Fortunato Bodziony, F., Philippe, R., Fongarland, P., & Augier, F. (2022). Use of CFD for pressure drop, liquid saturation and wetting predictions in trickle bed reactors for different catalyst particle shapes. *Chemical Engineering Science, 249*, 117315. https://doi.org/10.1016/j.ces.2021.117315

Brannock, M., Wang, Y., & Leslie, G. (2010). Mixing characterisation of full-scale membrane bioreactors: CFD modelling with experimental validation. *Water Research, 44*(10), 3181–3191. https://doi.org/10.1016/j.watres.2010.02.029

Camp, T. R. (1942). Grit chamber design. In *Source: Sewage Works Journal* (Vol. 14, Issue 2). https://about.jstor.org/terms

Cantarero Rivera, F. J., & Chen, J. (2022). Computational fluid dynamics modeling of cell cultures in bioreactors and its potential for cultivated meat production—A mini-review. *Future Foods, 6*, 100195. https://doi.org/10.1016/j.fufo.2022.100195

Carey, R. O., & Migliaccio, K. W. (2009). Contribution of wastewater treatment plant effluents to nutrient dynamics in aquatic systems. In *Environmental Management* (Vol. 44, Issue 2, pp. 205–217). https://doi.org/10.1007/s00267-009-9309-5

Carpentieri, M., Kumar, P., & Robins, A. (2011). An overview of experimental results and dispersion modelling of nanoparticles in the wake of moving vehicles. *Environmental Pollution, 159*(3), 685–693. https://doi.org/10.1016/j.envpol.2010.11.041

Casula, E., Molognoni, D., Borràs, E., & Mascia, M. (2021). 3D modelling of bioelectrochemical systems with brush anodes under fed-batch and flow conditions. *Journal of Power Sources, 487*, 229432. https://doi.org/10.1016/j.jpowsour.2020.229432

Chang, J., Zhang, K., Chen, H., Yang, Y., & Zhang, L. (2013). CFD modelling of the hydrodynamics and kinetic reactions in a fluidised-bed MTO reactor. *Chemical Engineering Research and Design, 91*(12), 2355–2368. https://doi.org/10.1016/J.CHERD.2013.04.023

Chen, B., Wei, G., Zhang, T., Wu, H., Wu, C., Chen, A., Zhang, H., Guan, X., Ren, Y., Feng, C., Qiu, G., & Wei, C. (2023). Flow drag force contributes high bio-treatment efficiency in a circulating fluidized bed reactor: Mechanism of selective separation of functional decayed sludge. *Chemical Engineering Journal, 454*, 140448. https://doi.org/10.1016/j.cej.2022.140448

Climent, J., Basiero, L., Martínez-Cuenca, R., Berlanga, J. G., Julián-López, B., & Chiva, S. (2018). Biological reactor retrofitting using CFD-ASM modelling. *Chemical Engineering Journal, 348*, 1–14. https://doi.org/10.1016/j.cej.2018.04.058

Cornelissen, J. T., Taghipour, F., Escudié, R., Ellis, N., & Grace, J. R. (2007). CFD modelling of a liquid–solid fluidized bed. *Chemical Engineering Science, 62*(22), 6334–6348. https://doi.org/10.1016/j.ces.2007.07.014

D'Bastiani, C., Kennedy, D., & Reynolds, A. (2023). CFD simulation of anaerobic granular sludge reactors: A review. *Water Research, 242*, 120220. https://doi.org/10.1016/j.watres.2023.120220

Dabiri, S., Kumar, P., & Rauch, W. (2023). Integrating biokinetics with computational fluid dynamics for energy performance analysis in anaerobic digestion. *Bioresource Technology, 373*, 128728. https://doi.org/10.1016/j.biortech.2023.128728

Dai, Z., Yu, R., Wu, Y., Zhu, G., Lu, X., & Zha, X. (2022). Modelling of self-sustainable microbial fuel cell type oil sensors based on restricted oxygen transfer and two-population competition. *Science of the Total Environment, 806*, 151333. https://doi.org/10.1016/j.scitotenv.2021.151333

Das, S., Bai, H., Wu, C., Kao, J. H., Barney, B., Kidd, M., & Kuettel, M. (2016). Improving the performance of industrial clarifiers using three-dimensional computational fluid dynamics. *Engineering Applications of Computational Fluid Mechanics, 10*(1), 130–144. https://doi.org/10.1080/19942060.2015.1121518

Deb, D., Patel, R., & Balas, V. E. (2020). A review of control-oriented bioelectrochemical mathematical models of microbial fuel cells. *Processes, 8*(5), 583. https://doi.org/10.3390/pr8050583

Dehnavi, M. A., Shahhosseini, S., Hashemabadi, S. H., & Ghafelebashi, S. M. (2010). CFD simulation of hydrodynamics and heat transfer in gas phase ethylene polymerization reactors. *International Communications in Heat and Mass Transfer, 37*(4), 437–442. https://doi.org/10.1016/j.icheatmasstransfer.2009.12.005

Dick, E. (n.d.). Introduction to finite volume methods in computational fluid dynamics. In *Computational fluid dynamics* (pp. 275–301). Springer, Berlin. https://doi.org/10.1007/978-3-540-85056-4_11

Dick, E. (2009). Introduction to finite element methods in computational fluid dynamics. In *Computational fluid dynamics* (pp. 235–274). Springer, Berlin. https://doi.org/10.1007/978-3-540-85056-4_10

Dixon, A. G., & Partopour, B. (2020). Computational fluid dynamics for fixed bed reactor design. *Annual Review of Chemical and Biomolecular Engineering, 11*(1), 109–130. https://doi.org/10.1146/annurev-chembioeng-092319-075328

Dong, Y., Sosna, B., Korup, O., Rosowski, F., & Horn, R. (2017). Investigation of radial heat transfer in a fixed-bed reactor: CFD simulations and profile measurements. *Chemical Engineering Journal, 317*, 204–214. https://doi.org/10.1016/j.cej.2017.02.063

Doumic, L. I., Soares, P. A., Ayude, M. A., Cassanello, M., Boaventura, R. A. R., & Vilar, V. J. P. (2015). Enhancement of a solar photo-Fenton reaction by using ferrioxalate complexes for the treatment of a synthetic cotton-textile dyeing wastewater. *Chemical Engineering Journal, 277*, 86–96. https://doi.org/10.1016/j.cej.2015.04.074

Dutta, A., Das, N., Sarkar, D., & Chakrabarti, S. (2019). Development and characterization of a continuous solar-collector-reactor for wastewater treatment by photo-Fenton process. *Solar Energy, 177*, 364–373. https://doi.org/10.1016/j.solener.2018.11.036

Ed Fontes. (2018, November 29). FEM vs. FVM. COMSOL.

Eshtiaghi, N., Markis, F., Yap, S. D., Baudez, J. C., & Slatter, P. (2013). Rheological characterisation of municipal sludge: A review. In *Water Research* (Vol. 47, Issue 15, pp. 5493–5510). Elsevier Ltd. https://doi.org/10.1016/j.watres.2013.07.001

Falcão, D. M. (1997). *High performance computing in power system applications* (pp. 1–23). https://doi.org/10.1007/3-540-62828-2_109

Farber, P., Gräbel, J., Kroppen, N., Pötschke, L., Roos, D., Rosenbaum, M., Stegschuster, G., & Ueberholz, P. (2021). Electricity generation in a microbial fuel cell with textile carbon fibre anodes. *Computers & Mathematics with Applications, 83*, 4–23. https://doi.org/10.1016/j.camwa.2019.11.019

Farzan, P., Mistry, B., & Ierapetritou, M. G. (2017). Review of the important challenges and opportunities related to modeling of mammalian cell bioreactors. *AIChE Journal, 63*(2), 398–408. https://doi.org/10.1002/aic.15442

Feickert Fenske, C., Strübing, D., & Koch, K. (2023). Biological methanation in trickle bed reactors—a critical review. *Bioresource Technology, 385*, 129383. https://doi.org/10.1016/j.biortech.2023.129383

Fu, B., Xu, T., Guo, X., Liang, P., Huang, X., & Zhang, X. (2019). Optimization and simulation of a carbon-based flow-through composite anode configuration to enhance power generation and improve effluent quality simultaneously for microbial fuel cells. *Journal of Cleaner Production, 229*, 542–551. https://doi.org/10.1016/j.jclepro.2019.04.308

Karthik, G. M., & Buwa, V. V. (2018). Particle-resolved simulations of methane steam reforming in multilayered packed beds. *AIChE Journal, 64*(11), 4162–4176. https://doi.org/10.1002/aic.16386

Gadkari, S., Gu, S., & Sadhukhan, J. (2018). Towards automated design of bioelectrochemical systems: A comprehensive review of mathematical models. *Chemical Engineering Journal, 343*, 303–316. https://doi.org/10.1016/j.cej.2018.03.005

Gao, H., & Stenstrom, M. K. (n.d.). *World environmental and water resources congress 2020.*

Gao, H., & Stenstrom, M. K. (2018). Evaluation of three turbulence models in predicting the steady state hydrodynamics of a secondary sedimentation tank. Water Research, 143, 445–456. https://doi.org/10.1016/j.watres.2018.06.067

Gao, H., & Stenstrom, M. K. (2020). Development and applications in computational fluid dynamics modeling for secondary settling tanks over the last three decades: A review. In *Water Environment Research* (Vol. 92, Issue 6, pp. 796–820). Wiley, New York. https://doi.org/10.1002/wer.1279

Glessmer, M., & Janßen, C. (2017). Using an interactive Lattice Boltzmann Solver in fluid mechanics instruction. *Computation, 5*(4), 35. https://doi.org/10.3390/computation5030035

Gómez-Ríos, D., Junne, S., Neubauer, P., Ochoa, S., Ríos-Estepa, R., & Ramírez-Malule, H. (2019). Characterization of the metabolic response of Streptomyces clavuligerus to shear stress in stirred tanks and single-use 2D rocking motion bioreactors for clavulanic acid production. *Antibiotics, 8*(4), 168. https://doi.org/10.3390/antibiotics8040168

Goswami, L., Vinoth Kumar, R., Borah, S. N., Arul Manikandan, N., Pakshirajan, K., & Pugazhenthi, G. (2018). Membrane bioreactor and integrated membrane bioreactor systems for micropollutant removal from wastewater: A review. *Journal of Water Process Engineering, 26*, 314–328. https://doi.org/10.1016/j.jwpe.2018.10.024

Gu, Y., Li, Y., Yuan, F., & Yang, Q. (2023). Optimization and control strategies of aeration in WWTPs: A review. *Journal of Cleaner Production*. https://doi.org/10.1016/j.jclepro.2023.138008

Guzmán, C., Carrera, J., Durán, H., Berumen, J., Ortiz, A., Guirette, O., Arroyo, A., Brizuela, J., Gómez, F., Blanco, A., Azcaray, H., & Hernández, M. (2018). Implementation of virtual sensors for monitoring temperature in greenhouses using CFD and control. *Sensors, 19*(1), 60. https://doi.org/10.3390/s19010060

Hassanzadeh, A., Firouzi, M., Albijanic, B., & Celik, M. S. (2018). A review on determination of particle–bubble encounter using analytical, experimental and numerical methods. *Minerals Engineering, 122*, 296–311. https://doi.org/10.1016/j.mineng.2018.04.014

He, W., Li, Z. K., Tang, A. R., Liu, A., & He, B. G. (2013). Simulation study of three-phase flow field based on microbubble flotation. *Applied Mechanics and Materials, 477–478*, 226–231. https://doi.org/10.4028/www.scientific.net/AMM.477-478.226

Heroux, M. A. (2022). Research software science: Expanding the impact of research software engineering. *Computing in Science & Engineering, 24*(6), 22–27. https://doi.org/10.1109/MCSE.2023.3260475

Hickman, D. A., Degenstein, J. C., & Ribeiro, F. H. (2016). Fundamental principles of laboratory fixed bed reactor design. *Current Opinion in Chemical Engineering, 13*, 1–9. https://doi.org/10.1016/j.coche.2016.07.002

Hoiberg, B., & Shah, M. T. (2021). CFD study of multiphase flow in aerated grit tank. *Journal of Water Process Engineering*. https://doi.org/10.1016/j.jwpe.2020.101698

Hosseini, S. H., Ahmadi, G., & Olazar, M. (2013). CFD simulation of cylindrical spouted beds by the kinetic theory of granular flow. *Powder Technology, 246*, 303–316. https://doi.org/10.1016/j.powtec.2013.05.016

Hosseini, S. H., Rahimi, R., Zivdar, M., & Samimi, A. (2009). The effect of ring baffles on the hydrodynamics of a gas—solid bubbling fluidized bed using computational fluid dynamics. *Proceedings of the Institution of Mechanical Engineers, Part C: Journal of Mechanical Engineering Science, 223*(10), 2281–2289. https://doi.org/10.1243/09544062JMES1575

Hreiz, R., Potier, O., Wicks, J., & Commenge, J. M. (2019). CFD Investigation of the effects of bubble aerator layouts on hydrodynamics of an activated sludge channel reactor. *Environmental Technology (united Kingdom), 40*(20), 2657–2670. https://doi.org/10.1080/09593330.2018.1448001

https://www.cfdsupport.com/hardware-for-cfd.html. (n.d.). *Hardware for CFD simulations.*

https://www.comsol.com/. (n.d.). *COMSOL Multiphysics® simulation software understand, predict, and optimize real-world designs, devices, and processes with simulation.*

Huang, Q., & Long, X. (2020). Analysis of the Influencing Factors on Oil Removal Efficiency in Large-Scale Flotation Tanks: Experimental Observation and Numerical Simulation. Energies, 13(4), 927. https://doi.org/10.3390/en13040927

Hyre, M. (2012). *Evaluation of three full-scale grit removal processes using CFD modeling*. https://www.researchgate.net/publication/265786513

Itoh, N., & Mimura, K. (2013). Computational fluid dynamics (CFD) analysis of membrane reactors: Simulation of single-and multi-tube palladium membrane reactors for hydrogen recovery from cyclohexane. In *Handbook of Membrane Reactors* (Vol. 1, pp. 464–495). Elsevier Inc. https://doi.org/10.1533/9780857097330.3.464

Jafarinejad, S. (2017). Cost estimation and economical evaluation of three configurations of activated sludge process for a wastewater treatment plant (WWTP) using simulation. *Applied Water Science, 7*(5), 2513–2521. https://doi.org/10.1007/S13201-016-0446-8

Jalilnejad, E., Jabbari, B., & Ghasemzadeh, K. (2022). Application of computational fluid dynamics technique in membrane bioreactor systems. In *Current Trends and Future Developments on (Bio-) Membranes* (pp. 345–375). Elsevier. https://doi.org/10.1016/B978-0-12-822294-2.00007-2

Jamshidian, R., Scully, J., & Van den Akker, H. E. A. (2023). Two-fluid simulations of an aerated lab-scale bioreactor. *Chemical Engineering Research and Design, 196*, 254–275. https://doi.org/10.1016/j.cherd.2023.06.042

Jasak, H. (2009). OpenFOAM: Open source CFD in research and industry. *International Journal of Naval Architecture and Ocean Engineering, 1*(2), 89–94. https://doi.org/10.2478/IJNAOE-2013-0011

Jasak, H., & Tuković. (2010). *Dynamic mesh handling in openfoam applied to fluid-structure interaction simulations.*

Jasak, H., & Uroić, T. (2020). *Practical computational fluid dynamics with the finite volume method* (pp. 103–161). CISM International Centre for Mechanical Sciences. https://doi.org/10.1007/978-3-030-37518-8_4

Jeong, W., & Seong, J. (2014). Comparison of effects on technical variances of computational fluid dynamics (CFD) software based on finite element and finite volume methods. *International Journal of Mechanical Sciences, 78*, 19–26. https://doi.org/10.1016/j.ijmecsci.2013.10.017

Jin, Y., Liu, C. L., Song, X. F., & Yu, J. G. (2019). Computational fluid dynamics simulation as a tool for optimizing the hydrodynamic performance of membrane bioreactors. *RSC Advances, 9*(55), 32034–32046. https://doi.org/10.1039/c9ra06706j

John, D. (n.d.). *Improved design of settling tanks using an extended drift flux model.* http://research-information.bristol.ac.uk

Jurtz, N., Kraume, M., & Wehinger, G. D. (2019). Advances in fixed-bed reactor modeling using particle-resolved computational fluid dynamics (CFD). *Reviews in Chemical Engineering, 35*(2), 139–190. https://doi.org/10.1515/revce-2017-0059

Kamyar, A., Saidur, R., & Hasanuzzaman, M. (2012). Application of computational fluid dynamics (CFD) for nanofluids. *International Journal of Heat and Mass Transfer, 55*(15–16), 4104–4115. https://doi.org/10.1016/j.ijheatmasstransfer.2012.03.052

Karn, A. L., Pandya, S., Mehbodniya, A., Arslan, F., Sharma, D. K., Phasinam, K., Aftab, M. N., Rajan, R., Bommisetti, R. K., & Sengan, S. (2023). An integrated approach for sustainable development of wastewater treatment and management system using IoT in smart cities. *Soft Computing, 27*(8), 5159–5175. https://doi.org/10.1007/s00500-021-06244-9

Karpinska, A. M., & Bridgeman, J. (2016a). CFD-aided modelling of activated sludge systems—A critical review. *Water Research, 88*, 861–879. https://doi.org/10.1016/j.watres.2015.11.008

Karpinska, A. M., & Bridgeman, J. (2016b). CFD-aided modelling of activated sludge systems—A critical review. In *Water Research* (Vol. 88, pp. 861–879). Elsevier Ltd. https://doi.org/10.1016/j.watres.2015.11.008

Kim, J., Kim, H., Kim, B., & Yu, J. (2014). Computational fluid dynamics analysis in microbial fuel cells with different anode configurations. *Water Science and Technology, 69*(7), 1447–1452. https://doi.org/10.2166/wst.2014.041

Koerich, D. M., Lopes, G. C., & Rosa, L. M. (2018). Investigation of phases interactions and modification of drag models for liquid-solid fluidized bed tapered bioreactors. *Powder Technology, 339*, 90–101. https://doi.org/10.1016/j.powtec.2018.07.102

Kone, J.-P., Zhang, X., Yan, Y., Hu, G., & Ahmadi, G. (2017). Three-dimensional multiphase flow computational fluid dynamics models for proton exchange membrane fuel cell: A theoretical

development. *The Journal of Computational Multiphase Flows, 9*(1), 3–25. https://doi.org/10.1177/1757482X17692341
Korzani, M. G., Galindo Torres, S., Scheuermann, A., & Williams, D. J. (2016). Smoothed particle hydrodynamics into the fluid dynamics of classical problems. *Applied Mechanics and Materials, 846*, 73–78. https://doi.org/10.4028/www.scientific.net/AMM.846.73
Kostoglou, M., Karapantsios, T. D., & Matis, K. A. (2007). CFD model for the design of large scale flotation tanks for water and wastewater treatment. *Industrial and Engineering Chemistry Research, 46*(20), 6590–6599. https://doi.org/10.1021/ie0703989
Kumar, T., & Eswari, J. S. (2023). Review and perspectives of emerging green technology for the sequestration of carbon dioxide into value-added products: An intensifying development. *Energy & Fuels, 37*(5), 3570–3589. https://doi.org/10.1021/acs.energyfuels.2c04122
Kumar, T., & Eswari Jujjavarappu, S. (2023). A critical review on an advanced bio-electrochemical system for carbon dioxide sequestration and wastewater treatment. *Total Environment Research Themes, 5*, 100023. https://doi.org/10.1016/j.totert.2022.100023
Kumar, T., & Jujjavarapu, S. E. (2023a). Advances in anode configurations for a microbial fuel cell via a computational fluid dynamics electrochemistry and its experimental validation. *Journal of Chemical Technology & Biotechnology*. https://doi.org/10.1002/jctb.7397
Kumar, T., & Jujjavarapu, S. E. (2023b). Carbon dioxide sequestration and wastewater treatment via an innovative self-sustaining algal microbial fuel cell. *Journal of Cleaner Production, 415*, 137836. https://doi.org/10.1016/j.jclepro.2023.137836
Kumar, T., Naik, S., & Jujjavarappu, S. E. (2021). A critical review on early-warning electro chemical system on microbial fuel cell-based biosensor for on-site water quality monitoring. *Chemosphere*. https://doi.org/10.1016/j.chemosphere.2021.133098
Kumar, T., Naik, S., & Jujjavarappu, S. E. (2022). A critical review on early-warning electrochemical system on microbial fuel cell-based biosensor for on-site water quality monitoring. *Chemosphere, 291*, 133098. https://doi.org/10.1016/j.chemosphere.2021.133098
Kwon, S. B., Park, N. S., Lee, S. J., Ahn, H. W., & Wang, C. K. (2006). Examining the effect of length/width ratio on the hydro-dynamic behaviour in a DAF system using CFD and ADV techniques. *Water Science and Technology, 53*(7), 141–149. https://doi.org/10.2166/wst.2006.218
Lamon, L., Asturiol, D., Vilchez, A., Cabellos, J., Damásio, J., Janer, G., Richarz, A., & Worth, A. (2019). Physiologically based mathematical models of nanomaterials for regulatory toxicology: A review. *Computational Toxicology, 9*, 133–142. https://doi.org/10.1016/j.comtox.2018.10.002
le Moullec, Y., Gentric, C., Potier, O., & Leclerc, J. P. (2010). CFD simulation of the hydrodynamics and reactions in an activated sludge channel reactor of wastewater treatment. *Chemical Engineering Science, 65*(1), 492–498. https://doi.org/10.1016/j.ces.2009.03.021
Lee, I.-B., Bitog, J. P. P., Hong, S.-W., Seo, I.-H., Kwon, K.-S., Bartzanas, T., & Kacira, M. (2013). The past, present and future of CFD for agro-environmental applications. *Computers and Electronics in Agriculture, 93*, 168–183. https://doi.org/10.1016/j.compag.2012.09.006
Lee, K. H., Kim, H., KuK, J. W., Chung, J. D., Park, S., & Kwon, E. E. (2020). Micro-bubble flow simulation of dissolved air flotation process for water treatment using computational fluid dynamics technique. *Environmental Pollution*. https://doi.org/10.1016/j.envpol.2019.01.011
Lee, J. L., & Lim, E. W. C. (2017). Comparisons of Eulerian-Eulerian and CFD-DEM simulations of mixing behaviors in bubbling fluidized beds. *Powder Technology, 318*, 193–205. https://doi.org/10.1016/j.powtec.2017.05.050
Lettieri, P., & Mazzei, L. (2009). Challenges and issues on the CFD modeling of fluidized beds: A review. *The Journal of Computational Multiphase Flows, 1*(2), 83–131. https://doi.org/10.1260/175748209789563937
Li, L., Wang, K., Zhao, Q., Gao, Q., Zhou, H., Jiang, J., & Mei, W. (2022). A critical review of experimental and CFD techniques to characterize the mixing performance of anaerobic digesters for biogas production. *Reviews in Environmental Science and Bio/technology, 21*(3), 665–689. https://doi.org/10.1007/s11157-022-09626-z
Li, W., Li, M., Bolton, J. R., & Qiang, Z. (2016). Configuration optimization of UV reactors for water disinfection with computational fluid dynamics: Feasibility of using particle minimum

UV dose as a performance indicator. *Chemical Engineering Journal, 306*, 1–8. https://doi.org/10.1016/j.cej.2016.07.042

Li, W., Li, M., Wen, D., & Qiang, Z. (2018a). Development of economical-running strategy for multi-lamp UV disinfection reactors in secondary water supply systems with computational fluid dynamics simulations. *Chemical Engineering Journal, 343*, 317–323. https://doi.org/10.1016/j.cej.2018.03.017

Li, X., Scott, K., Kelly, W. J., & Huang, Z. (2018b). Development of a computational fluid dynamics model for scaling-up Ambr bioreactors. *Biotechnology and Bioprocess Engineering, 23*(6), 710–725. https://doi.org/10.1007/s12257-018-0063-5

Lim, S. S., Fontmorin, J.-M., Pham, H. T., Milner, E., Abdul, P. M., Scott, K., Head, I., & Yu, E. H. (2021). Zinc removal and recovery from industrial wastewater with a microbial fuel cell: Experimental investigation and theoretical prediction. *Science of the Total Environment, 776*, 145934. https://doi.org/10.1016/j.scitotenv.2021.145934

Lin, J.-R., Cao, J., Zhang, J.-P., van Treeck, C., & Frisch, J. (2019). Visualization of indoor thermal environment on mobile devices based on augmented reality and computational fluid dynamics. *Automation in Construction, 103*, 26–40. https://doi.org/10.1016/j.autcon.2019.02.007

Liu, X., Wang, Y., Shi, Y., Li, Q., Dai, P., Guan, J., Waite, T. D., & Leslie, G. (2019). CFD modelling of uneven flows behaviour in flat-sheet membrane bioreactors: From bubble generation to shear stress distribution. *Journal of Membrane Science, 570–571*, 146–155. https://doi.org/10.1016/j.memsci.2018.10.040

Liu, X., Wang, Y., Waite, T. D., & Leslie, G. (2015). Numerical simulation of bubble induced shear in membrane bioreactors: Effects of mixed liquor rheology and membrane configuration. *Water Research, 75*, 131–145. https://doi.org/10.1016/j.watres.2015.02.009

Llano-Serna, C. E., Coral-Portillo, J. A., Fontalvo, J., & Prado-Rubio, O. A. (2019). CFD assisted analysis and design of hydraulic flocculators. *Revista Mexicana de Ingeniera Quimica, 18*(3), 995–1015. https://doi.org/10.24275/UAM/IZT/DCBI/REVMEXINGQUIM/2019V18N3/LLANO

Logtenberg, S. A., & Dixon, A. G. (1998). Computational fluid dynamics studies of fixed bed heat transfer. *Chemical Engineering and Processing: Process Intensification, 37*(1), 7–21. https://doi.org/10.1016/S0255-2701(97)00032-9

Lopes, R. J. G., & Quinta-Ferreira, R. M. (2007). Trickle-bed CFD studies in the catalytic wet oxidation of phenolic acids. *Chemical Engineering Science, 62*(24), 7045–7052. https://doi.org/10.1016/j.ces.2007.08.085

Lopes, R. J. G., Silva, A. M. T., & Quinta-Ferreira, R. M. (2007). Kinetic modeling and trickle-bed CFD studies in the catalytic wet oxidation of Vanillic acid. *Industrial & Engineering Chemistry Research, 46*(25), 8380–8387. https://doi.org/10.1021/ie070009a

López-Jiménez, P. A., Escudero-González, J., Montoya Martínez, T., Fajardo Montañana, V., & Gualtieri, C. (2015). Application of CFD methods to an anaerobic digester: The case of Ontinyent WWTP, Valencia, Spain. *Journal of Water Process Engineering, 7*, 131–140. https://doi.org/10.1016/j.jwpe.2015.05.006

Luo, S., Sun, H., Ping, Q., Jin, R., & He, Z. (2016). A review of modeling bioelectrochemical systems: Engineering and statistical aspects. *Energies, 9*(2), 111. https://doi.org/10.3390/en9020111

Macfarlan, L. H., Phan, M. T., & Bruce Eldridge, R. (2021). Structured packing geometry study for liquid-phase mass transfer and hydrodynamic performance using CFD. *Chemical Engineering Science*. https://doi.org/10.1016/J.CES.2021.117353

Makatsa, T. J., Baloyi, J., Ntho, T., & Masuku, C. M. (2021a). Catalytic wet air oxidation of phenol: Review of the reaction mechanism, kinetics, and CFD modeling. *Critical Reviews in Environmental Science and Technology, 51*(17), 1891–1923. https://doi.org/10.1080/10643389.2020.1771886

Makatsa, T. J., Baloyi, S. J., & Masuku, C. M. (2021). *Computational fluid dynamics modelling of phenol oxidation in a trickle-bed reactor using 3D Eulerian model* (pp. 825–831). https://doi.org/10.1016/B978-0-323-88506-5.50129-7

Malang, J., Kumar, P., & Saptoro, A. (2015). Computational fluid dynamics-based hydrodynamics studies in packed bed columns: Current status and future directions. *International Journal of Chemical Reactor Engineering, 13*(3), 289–303. https://doi.org/10.1515/ijcre-2014-0121

Malato, S., Fernández-Ibáñez, P., Maldonado, M. I., Blanco, J., & Gernjak, W. (2009). Decontamination and disinfection of water by solar photocatalysis: Recent overview and trends. In *Catalysis Today* (Vol. 147, Issue 1, pp. 1–59). https://doi.org/10.1016/j.cattod.2009.06.018

Mandenius, C.-F. (2016). Challenges for bioreactor design and operation. In *Bioreactors* (pp. 1–34). Wiley-VCH Verlag GmbH & Co. KGaA. https://doi.org/10.1002/9783527683369.ch1

Massaglia, G., Gerosa, M., Agostino, V., Cingolani, A., Sacco, A., Saracco, G., Margaria, V., & Quaglio, M. (2017). Fluid dynamic modeling for microbial fuel cell based biosensor optimization. *Fuel Cells, 17*(5), 627–634. https://doi.org/10.1002/fuce.201700026

Matko, T., Chew, J., Wenk, J., Chang, J., & Hofman, J. (2021). Computational fluid dynamics simulation of two-phase flow and dissolved oxygen in a wastewater treatment oxidation ditch. *Process Safety and Environmental Protection, 145*, 340–353. https://doi.org/10.1016/j.psep.2020.08.017

Mehl, M., Neckel, T., & Neumann, P. H. (2011). Navier-Stokes and Lattice-Boltzmann on octree-like grids in the Peano framework. *International Journal for Numerical Methods in Fluids, 65*(1–3), 67–86. https://doi.org/10.1002/fld.2469

Meroney, R. N., & Sheker, R. E. (2015). Removing grit during wastewater treatment: CFD analysis of HDVS performance. *Water Environment Research, 88*(5), 438–448. https://doi.org/10.2175/106143015x14212658614478

Mihalič, F., Truntič, M., & Hren, A. (2022). Hardware-in-the-loop simulations: A historical overview of engineering challenges. *Electronics, 11*(15), 2462. https://doi.org/10.3390/electronics11152462

Minin, O. (Ed.). (2011). *Computational fluid dynamics technologies and applications*. InTech. https://doi.org/10.5772/686

Morchain, J., Gabelle, J.-C., & Cockx, A. (2013). Coupling of biokinetic and population balance models to account for biological heterogeneity in bioreactors. *AIChE Journal, 59*(2), 369–379. https://doi.org/10.1002/aic.13820

Moreira, F. C., Boaventura, R. A. R., Brillas, E., & Vilar, V. J. P. (2017). Electrochemical advanced oxidation processes: A review on their application to synthetic and real wastewaters. In *Applied Catalysis B: Environmental* (Vol. 202, pp. 217–261). Elsevier B.V. https://doi.org/10.1016/j.apcatb.2016.08.037

Moustakas, K., Rehan, M., Loizidou, M., Nizami, A. S., & Naqvi, M. (2020). Energy and resource recovery through integrated sustainable waste management. *Applied Energy, 261*, 114372. https://doi.org/10.1016/j.apenergy.2019.114372

Mrope, H. A., Chande Jande, Y. A., & Kivevele, T. T. (2021). A review on computational fluid dynamics applications in the design and optimization of crossflow hydro turbines. *Journal of Renewable Energy, 2021*, 1–13. https://doi.org/10.1155/2021/5570848

Muhammad, N. (2021). Finite volume method for simulation of flowing fluid via OpenFOAM. *The European Physical Journal plus, 136*(10), 1010. https://doi.org/10.1140/epjp/s13360-021-01983-y

Nadal-Rey, G., McClure, D. D., Kavanagh, J. M., Cornelissen, S., Fletcher, D. F., & Gernaey, K. V. (2021). Understanding gradients in industrial bioreactors. *Biotechnology Advances, 46*, 107660. https://doi.org/10.1016/j.biotechadv.2020.107660

Naik, S., & Jujjavarappu, S. E. (2020). Simultaneous bioelectricity generation from cost-effective MFC and water treatment using various wastewater samples. *Environmental Science and Pollution Research, 27*(22), 27383–27393. https://doi.org/10.1007/s11356-019-06221-8

Nakatsuka, N., Kishita, Y., Kurafuchi, T., & Akamatsu, F. (2020). Integrating wastewater treatment and incineration plants for energy-efficient urban biomass utilization: A life cycle analysis. *Journal of Cleaner Production, 243*, 118448. https://doi.org/10.1016/J.JCLEPRO.2019.118448

Norouzi-Firouz, H., Sarrafzadeh, M., Zarghami, R., & Moshiri-Tabrizi, I. (2022). A coupled hydrodynamic-biokinetic simulation of three-phase flow in an oxidation ditch using CFD.

The Canadian Journal of Chemical Engineering, 100(2), 223–236. https://doi.org/10.1002/cjce.24095

Oh, S. T. (2010). *Trend of mathematical models in microbial fuel cell for environmental energy refinery from waste/water* (pp. 25–30). https://doi.org/10.1007/978-3-642-13624-5_3

Ortiz-Martínez, V. M., Salar-García, M. J., de los Ríos, A. P., Hernández-Fernández, F. J., Egea, J. A., & Lozano, L. J. (2015). Developments in microbial fuel cell modeling. *Chemical Engineering Journal, 271*, 50–60. https://doi.org/10.1016/j.cej.2015.02.076

Ortiz-Martínez, V. M., Salar-García, M. J., los Ríos, A. P., Hernández-Fernández, F. J., Sánchez-Segado, S., Egea-Larrosa, J. A., & Lozano-Blanco, L. J. (2014). Avances recientes en modelado y simulación de pilas de combustible microbianas. *DYNA-Ingeniería e Industria, 89*(6).

Ou, S., Kashima, H., Aaron, D. S., Regan, J. M., & Mench, M. M. (2017). Full cell simulation and the evaluation of the buffer system on air-cathode microbial fuel cell. *Journal of Power Sources, 347*, 159–169. https://doi.org/10.1016/j.jpowsour.2017.02.031

Owusu, P. A., & Asumadu-Sarkodie, S. (2016). A review of renewable energy sources, sustainability issues and climate change mitigation. *Cogent Engineering, 3*(1), 1167990. https://doi.org/10.1080/23311916.2016.1167990

Ozgun, H., Dereli, R. K., Ersahin, M. E., Kinaci, C., Spanjers, H., & van Lier, J. B. (2013). A review of anaerobic membrane bioreactors for municipal wastewater treatment: Integration options, limitations and expectations. *Separation and Purification Technology, 118*, 89–104. https://doi.org/10.1016/J.SEPPUR.2013.06.036

Palatsi, J., Ripoll, F., Benzal, A., Pijuan, M., & Romero-Güiza, M. S. (2021). Enhancement of biological nutrient removal process with advanced process control tools in full-scale wastewater treatment plant. *Water Research, 200*, 117212. https://doi.org/10.1016/j.watres.2021.117212

Pan, H., Chen, X.-Z., Liang, X.-F., Zhu, L.-T., & Luo, Z.-H. (2016). CFD simulations of gas–liquid–solid flow in fluidized bed reactors—A review. *Powder Technology, 299*, 235–258. https://doi.org/10.1016/j.powtec.2016.05.024

Panneerselvam, R., Savithri, S., & Surender, G. D. (2007). CFD based investigations on hydrodynamics and energy dissipation due to solid motion in liquid fluidised bed. *Chemical Engineering Journal, 132*(1–3), 159–171. https://doi.org/10.1016/j.cej.2007.01.042

Partopour, B., & Dixon, A. G. (2016). Reduced microkinetics model for computational fluid dynamics (CFD) simulation of the fixed-bed partial oxidation of ethylene. *Industrial & Engineering Chemistry Research, 55*(27), 7296–7306. https://doi.org/10.1021/acs.iecr.6b00526

Patziger, M. (2021). Improving wastewater treatment plant performance by applying CFD models for design and operation: Selected case studies. *Water Science and Technology, 84*(2), 323–332. https://doi.org/10.2166/wst.2021.019

Peleka, E. N., & Matis, K. A. (2016). Hydrodynamic aspects of flotation separation. *Open Chemistry, 14*(1), 132–139. https://doi.org/10.1515/chem-2016-0014

Pereira, J. P., Karpinska, A., Gomes, P. J., Martins, A. A., Dias, M. M., Lopes, J. C. B., & Santos, R. J. (2012). Activated sludge models coupled to CFD simulations. In *Single and two-phase flows on chemical and biomedical engineering* (pp. 153–173). Bentham Science Publishers Ltd. https://doi.org/10.2174/978160805295011201010153

Pilarek, M., Sobieszuk, P., Wierzchowski, K., & Dąbkowska, K. (2018). Impact of operating parameters on values of a volumetric mass transfer coefficient in a single-use bioreactor with wave-induced agitation. *Chemical Engineering Research and Design, 136*, 1–10. https://doi.org/10.1016/j.cherd.2018.04.012

Popat, S. C., & Torres, C. I. (2016). Critical transport rates that limit the performance of microbial electrochemistry technologies. *Bioresource Technology, 215*, 265–273. https://doi.org/10.1016/j.biortech.2016.04.136

Radeef, A. Y., & Ismail, Z. Z. (2019). Polarization model of microbial fuel cell for treatment of actual potato chips processing wastewater associated with power generation. *Journal of Electroanalytical Chemistry, 836*, 176–181. https://doi.org/10.1016/j.jelechem.2019.02.001

Ratkovich, N., Horn, W., Helmus, F. P., Rosenberger, S., Naessens, W., Nopens, I., & Bentzen, T. R. (2013). Activated sludge rheology: A critical review on data collection and modelling. *Water Research, 47*(2), 463–482. https://doi.org/10.1016/j.watres.2012.11.021

Rehman, U., Amerlinck, Y., Arnaldos, M., & Nopens, I. (n.d.). *CFD and Biokinetic model integration applied to a full scale WWTP.*

Ren, J., & Cao, S.-J. (2020). Development of self-adaptive low-dimension ventilation models using OpenFOAM: Towards the application of AI based on CFD data. *Building and Environment, 171*, 106671. https://doi.org/10.1016/j.buildenv.2020.106671

Ren, P., Li, W., & Yu, K. (2021). Computational fluid dynamics simulation of adsorption process in a liquid-solids fluidized bed. *Journal of Environmental Chemical Engineering, 9*(4), 105428. https://doi.org/10.1016/j.jece.2021.105428

Rivera-Alvarez, I., Brown, R. K., Keskin-Pyttel, D., Steffens, J., Farber, P., & Schröder, U. (2020). Correlating theoretical boundary layer thickness to the power output of a microbial fuel cell with a complex anode geometry operated at varying flow rates. *Journal of Power Sources, 470*, 228428. https://doi.org/10.1016/j.jpowsour.2020.228428

Runchal, A. K., & Rao, M. M. (2020). CFD of the future: Year 2025 and beyond. In *50 years of CFD in engineering sciences* (pp. 779–795). Springer Singapore. https://doi.org/10.1007/978-981-15-2670-1_22

Saini, A. K., Paritosh, K., Singh, A. K., & Vivekanand, V. (2021). CFD approach for pumped-recirculation mixing strategy in wastewater treatment: Minimizing power consumption, enhancing resource recovery in commercial anaerobic digester. *Journal of Water Process Engineering, 40*, 101777. https://doi.org/10.1016/j.jwpe.2020.101777

Salehpour, R., Jalilnejad, E., Nalband, M., & Ghasemzadeh, K. (2020). Hydrodynamic behavior of an airlift reactor with net draft tube with different configurations: Numerical evaluation using CFD technique. *Particuology, 51*, 91–108. https://doi.org/10.1016/j.partic.2019.09.005

Samstag, R. W., Ducoste, J. J., Griborio, A., Nopens, I., Batstone, D. J., Wicks, J. D., Saunders, S., Wicklein, E. A., Kenny, G., & Laurent, J. (2016a). CFD for wastewater treatment: An overview. *Water Science and Technology, 74*(3), 549–563. https://doi.org/10.2166/wst.2016.249

Samstag, R. W., Ducoste, J. J., Griborio, A., Nopens, I., Batstone, D. J., Wicks, J. D., Saunders, S., Wicklein, E. A., Kenny, G., & Laurent, J. (2016b). CFD for wastewater treatment: An overview. In *Water Science and Technology* (Vol. 74, Issue 3, pp. 549–563). IWA Publishing. https://doi.org/10.2166/wst.2016.249

Sánchez, F., Rey, H., Viedma, A., Nicolás-Pérez, F., Kaiser, A. S., & Martínez, M. (2018). CFD simulation of fluid dynamic and biokinetic processes within activated sludge reactors under intermittent aeration regime. *Water Research, 139*, 47–57. https://doi.org/10.1016/j.watres.2018.03.067

Sandhibigraha, S., Sasmal, S., Bandyopadhyay, T. K., & Bhunia, B. (2019). Computational fluid dynamics analysis of flow through immobilized catalyzed packed bed reactor for removal of 4-chlorophenol from wastewater. *Environmental Engineering Research, 25*(6), 878–889. https://doi.org/10.4491/eer.2019.184

Sangeetha, T., Li, I.-T., Lan, T.-H., Wang, C.-T., & Yan, W.-M. (2021). A fluid dynamics perspective on the flow dependent performance of honey comb microbial fuel cells. *Energy, 214*, 118928. https://doi.org/10.1016/j.energy.2020.118928

Sathe, S. M., Chakraborty, I., & Ghangrekar, M. M. (2021). Wastewater treatment and concomitant bioelectricity production using microbial fuel cell: Present aspects, up-scaling and future inventiveness. *Transactions of the Indian National Academy of Engineering, 6*(3), 633–651. https://doi.org/10.1007/s41403-021-00245-8

Satpathy, K., Rehman, U., Cools, B., Verdickt, L., Peleman, G., & Nopens, I. (2020). CFD-based process optimization of a dissolved air flotation system for drinking water production. *Water Science and Technology, 81*(8), 1668–1681. https://doi.org/10.2166/wst.2020.028

Schwedhelm, I., Zdzieblo, D., Appelt-Menzel, A., Berger, C., Schmitz, T., Schuldt, B., Franke, A., Müller, F.-J., Pless, O., Schwarz, T., Wiedemann, P., Walles, H., & Hansmann, J. (2019).

Automated real-time monitoring of human pluripotent stem cell aggregation in stirred tank reactors. *Scientific Reports, 9*(1), 12297. https://doi.org/10.1038/s41598-019-48814-w
Shadloo, M. S., Oger, G., & Le Touzé, D. (2016). Smoothed particle hydrodynamics method for fluid flows, towards industrial applications: Motivations, current state, and challenges. *Computers & Fluids, 136*, 11–34. https://doi.org/10.1016/j.compfluid.2016.05.029
Shen, L., Wu, Q., Ye, Q., Lin, H., Zhang, J., Chen, C., Yue, R., Teng, J., Hong, H., & Liao, B.-Q. (2023). Superior performance of a membrane bioreactor through innovative in-situ aeration and structural optimization using computational fluid dynamics. *Water Research*. https://doi.org/10.1016/j.watres.2023.120353
Shen, R., Jiao, Z., Parker, T., Sun, Y., & Wang, Q. (2020). Recent application of computational fluid dynamics (CFD) in process safety and loss prevention: A review. *Journal of Loss Prevention in the Process Industries, 67*, 104252. https://doi.org/10.1016/j.jlp.2020.104252
Shen, Z., Zhang, M., Fan, X., Shi, S., & Han, D. (2019). Hydrodynamic and flotation kinetic analysis of a large scale mechanical agitated flotation cell with the typical impeller and the arc impeller. *Minerals, 9*(2). https://doi.org/10.3390/min9020079
Shi, Y., Wang, Z., Du, X., Gong, B., Jegatheesan, V., & Haq, I. U. (2021). Recent advances in the prediction of fouling in membrane bioreactors. *Membranes, 11*(6), 381. https://doi.org/10.3390/membranes11060381
Shirazian, S., Rezakazemi, M., Marjani, A., & Moradi, S. (2012). Hydrodynamics and mass transfer simulation of wastewater treatment in membrane reactors. *Desalination, 286*, 290–295. https://doi.org/10.1016/j.desal.2011.11.039
Singh, H., & Hutmacher, D. W. (2009). *Bioreactor studies and computational fluid dynamics* (pp. 231–249). https://doi.org/10.1007/10_2008
Singhal, A., Cloete, S., Radl, S., Quinta-Ferreira, R., & Amini, S. (2017). Heat transfer to a gas from densely packed beds of monodisperse spherical particles. *Chemical Engineering Journal, 314*, 27–37. https://doi.org/10.1016/j.cej.2016.12.124
Sobieszuk, P., Zamojska-Jaroszewicz, A., & Makowski, Ł. (2017). Influence of the operational parameters on bioelectricity generation in continuous microbial fuel cell, experimental and computational fluid dynamics modelling. *Journal of Power Sources, 371*, 178–187. https://doi.org/10.1016/j.jpowsour.2017.10.032
Song, H.-S., Cannon, W., Beliaev, A., & Konopka, A. (2014). Mathematical modeling of microbial community dynamics: A methodological review. *Processes, 2*(4), 711–752. https://doi.org/10.3390/pr2040711
Song, X., Luo, W., Hai, F. I., Price, W. E., Guo, W., Ngo, H. H., & Nghiem, L. D. (2018). Resource recovery from wastewater by anaerobic membrane bioreactors: Opportunities and challenges. *Bioresource Technology, 270*, 669–677. https://doi.org/10.1016/j.biortech.2018.09.001
Suh, J.-W., Kim, J.-W., Choi, Y.-S., Kim, J.-H., Joo, W.-G., & Lee, K.-Y. (2018). Development of numerical Eulerian-Eulerian models for simulating multiphase pumps. *Journal of Petroleum Science and Engineering, 162*, 588–601. https://doi.org/10.1016/j.petrol.2017.10.073
Sultan, T., Ahmad, Z., Hayat, K., & Chaudhry, I. A. (2022). Computational analysis of three lamp close conduit water disinfection UV reactor. *International Journal of Environmental Science and Technology, 19*(5), 4393–4406. https://doi.org/10.1007/s13762-021-03344-9
Tan, J., Ji, Y.-N., Deng, W.-S., & Su, Y.-F. (2021). Process intensification in gas/liquid/solid reaction in trickle bed reactors: A review. *Petroleum Science, 18*(4), 1203–1218. https://doi.org/10.1016/j.petsci.2021.07.007
Trifi, D., Climent, J., Arnau, R., Carratalà, P., García, M., Beltrán, I., Badenes, C., Chiva, S., & Martínez-Cuenca, R. (2023). *Design and implementation of a passive agitator to increase UV dose in WWTPs disinfection channels* (pp. 626–636). https://doi.org/10.1007/978-3-031-15928-2_55
Tu, J., Yeoh, G. H., Liu, C., & Tao, Y. (2023). *Computational fluid dynamics: A practical approach*. Elsevier.
Udoewa, V., & Kumar, V. (2012). Computational fluid dynamics. *Applied Computational Fluid Dynamics*.

Wang, K., Li, L., Tan, F., & Wu, D. (2018). Treatment of landfill leachate using activated sludge technology: A review. *Archaea, 2018*, 1–10. https://doi.org/10.1155/2018/1039453

Wang, Y., Brannock, M., Cox, S., & Leslie, G. (2010). CFD simulations of membrane filtration zone in a submerged hollow fibre membrane bioreactor using a porous media approach. *Journal of Membrane Science, 363*(1–2), 57–66. https://doi.org/10.1016/j.memsci.2010.07.008

Wang, Y., Chen, J., & Larachi, F. (2013a). Modelling and simulation of trickle-bed reactors using computational fluid dynamics: A state-of-the-art review. *The Canadian Journal of Chemical Engineering, 91*(1), 136–180. https://doi.org/10.1002/cjce.20702

Wang, Y., Waite, T. D., & Leslie, G. L. (2013b). Computational fluid dynamics (CFD) analysis of membrane reactors: Modelling of membrane bioreactors for municipal wastewater treatment. In *Handbook of Membrane Reactors* (Vol. 1, pp. 532–568). Elsevier Inc. https://doi.org/10.1533/9780857097330.3.532

Werner, S., Kaiser, S. C., Kraume, M., & Eibl, D. (2014). Computational fluid dynamics as a modern tool for engineering characterization of bioreactors. *Pharmaceutical Bioprocessing, 2*(1), 85–99.

Weyand, B., Israelowitz, M., Kramer, J., Bodmer, C., Noehre, M., Strauss, S., Schmälzlin, E., Gille, C., von Schroeder, H. P., Reimers, K., & Vogt, P. M. (2015). Three-dimensional modelling inside a differential pressure laminar flow bioreactor filled with porous media. *BioMed Research International, 2015*, 1–9. https://doi.org/10.1155/2015/320280

Wicklein, E., Batstone, D. J., Ducoste, J., Laurent, J., Griborio, A., Wicks, J., Saunders, S., Samstag, R., Potier, O., & Nopens, I. (2016). Good modelling practice in applying computational fluid dynamics for WWTP modelling. *Water Science and Technology, 73*(5), 969–982. https://doi.org/10.2166/wst.2015.565

Wu, B. (2013). Advances in the use of CFD to characterize, design and optimize bioenergy systems. *Computers and Electronics in Agriculture, 93*, 195–208. https://doi.org/10.1016/j.compag.2012.05.008

Wu, B., & Chen, S. (2008). CFD simulation of non-Newtonian fluid flow in anaerobic digesters. *Biotechnology and Bioengineering, 99*(3), 700–711. https://doi.org/10.1002/bit.21613

Wu, Q., Yan, X., Xiao, K., Guan, J., Li, T., Liang, P., & Huang, X. (2018). Optimization of membrane unit location in a full-scale membrane bioreactor using computational fluid dynamics. *Bioresource Technology, 249*, 402–409. https://doi.org/10.1016/j.biortech.2017.09.209

Xanthos, S., Gong, M., Ramalingam, K., Fillos, J., Deur, A., Beckmann, K., & McCorquodale, J. A. (2011). Performance assessment of secondary settling tanks using CFD modeling. *Water Resources Management, 25*(4), 1169–1182. https://doi.org/10.1007/s11269-010-9620-1

Xia, C., Zhang, D., Pedrycz, W., Zhu, Y., & Guo, Y. (2018). Models for microbial fuel cells: A critical review. *Journal of Power Sources, 373*, 119–131. https://doi.org/10.1016/j.jpowsour.2017.11.001

Xiao, K., Liang, S., Wang, X., Chen, C., & Huang, X. (2019). Current state and challenges of full-scale membrane bioreactor applications: A critical review. *Bioresource Technology, 271*, 473–481. https://doi.org/10.1016/j.biortech.2018.09.061

Xiao, Y., Peng, J., Sun, W., Han, H., Xie, L., & Wang, R. (2023). Study on liquid-solid fluidisation behaviour of coarse coal particles: Experiment and CFD simulation. *Chemical Engineering Research and Design, 189*, 1–13. https://doi.org/10.1016/j.cherd.2022.11.001

Xie, F., Ge, H., Liu, J., Chen, W., & Song, H. (2018). CFD and experimental studies the effect of micro-channel turbulence promoter installation on the hydrodynamic performance of submerged flat-sheet membrane bioreactor. *Chemical Engineering and Processing - Process Intensification, 127*, 28–35. https://doi.org/10.1016/j.cep.2018.03.009

Xu, C., Rangaiah, G. P., & Zhao, X. S. (2015). A computational study of the effect of lamp arrangements on the performance of ultraviolet water disinfection reactors. *Chemical Engineering Science, 122*, 299–306. https://doi.org/10.1016/j.ces.2014.09.041

Xu, H., Yang, B., Liu, Y., Li, F., Shen, C., Ma, C., Tian, Q., Song, X., & Sand, W. (2018). Recent advances in anaerobic biological processes for textile printing and dyeing wastewater treatment: A mini-review. *World Journal of Microbiology and Biotechnology, 34*(11), 165. https://doi.org/10.1007/s11274-018-2548-y

Xu, N., Fan, L., Pang, H., & Shi, H. (2010). Feasibility study and CFD-aided design for a new type oxidation ditch based on airlift circulation. *The Canadian Journal of Chemical Engineering*. https://doi.org/10.1002/cjce.20335

Xu, Q., Wan, Y., Wu, Q., Xiao, K., Yu, W., Liang, S., Zhu, Y., Hou, H., Liu, B., Hu, J., Yang, Y., & Yang, J. (2021). An efficient hydrodynamic-biokinetic model for the optimization of operational strategy applied in a full-scale oxidation ditch by CFD integrated with ASM2. *Water Research, 193*, 116888. https://doi.org/10.1016/j.watres.2021.116888

Yang, J., Wu, J., Zhou, L., & Wang, Q. (2016). Computational study of fluid flow and heat transfer in composite packed beds of spheres with low tube to particle diameter ratio. *Nuclear Engineering and Design, 300*, 85–96. https://doi.org/10.1016/j.nucengdes.2015.10.030

Yao, Y., Fringer, O. B., & Criddle, C. S. (2022). CFD-accelerated bioreactor optimization: Reducing the hydrodynamic parameter space. *Environmental Science: Water Research & Technology, 8*(3), 456–464. https://doi.org/10.1039/D1EW00666E

Ye, C.-C., Zhang, P.-J.-Y., Wan, Z.-H., Yan, R., & Sun, D.-J. (2022). Accelerating CFD simulation with high order finite difference method on curvilinear coordinates for modern GPU clusters. *Advances in Aerodynamics, 4*(1), 7. https://doi.org/10.1186/s42774-021-00098-3

Yi, Y., Xie, B., Zhao, T., Qian, Z., & Liu, H. (2020). The effect of anode hydrodynamics on the sensitivity of microbial fuel cell based biosensors and the biological mechanism. *Bioelectrochemistry, 132*, 107351. https://doi.org/10.1016/j.bioelechem.2019.107351

Yoon, S.-H. (2015). *Membrane bioreactor processes: Principles and applications*. CRC Press.

Zawawi, M. H., Saleha, A., Salwa, A., Hassan, N. H., Zahari, N. M., Ramli, M. Z., & Muda, Z. C. (2018). *A review: Fundamentals of computational fluid dynamics (CFD)*. 020252. https://doi.org/10.1063/1.5066893

Zeng, Y., Choo, Y. F., Kim, B.-H., & Wu, P. (2010). Modelling and simulation of two-chamber microbial fuel cell. *Journal of Power Sources, 195*(1), 79–89. https://doi.org/10.1016/j.jpowsour.2009.06.101

Zhan, M., You, M., Liu, L., Zhang, Y., Yuan, F., Guo, B., Cheng, G., & Xu, W. (2021). Numerical simulation of mechanical flocculation in water treatment. *Journal of Environmental Chemical Engineering, 9*(4), 105536. https://doi.org/10.1016/J.JECE.2021.105536

Zhang, M., Dong, H., & Geng, Z. (2018). Computational study of flow and heat transfer in fixed beds with cylindrical particles for low tube to particle diameter ratios. *Chemical Engineering Research and Design, 132*, 149–161. https://doi.org/10.1016/j.cherd.2018.01.006

Zhang, S., Müller, D., Arellano-Garcia, H., & Wozny, G. (2013). CFD simulation of the fluid hydrodynamics in a continuous stirred-tank reactor. *Chemical Engineering Transactions, 32*. https://doi.org/10.3303/CET1332241

Zhao, L., Li, J., Battaglia, F., & He, Z. (2016). Investigation of multiphysics in tubular microbial fuel cells by coupled computational fluid dynamics with multi-order Butler-Volmer reactions. *Chemical Engineering Journal, 296*, 377–385. https://doi.org/10.1016/j.cej.2016.03.110

Chapter 3
An Overview of Computational Fluid Dynamics in Modelling and Simulation of Microbial Fuel Cells

1 Introduction

Microbial fuel cells (MFCs) continue to garner significant research attention as promising carbon-free and clean alternative energy sources. Nevertheless, optimizing their power output and organic content reduction in the laboratory remains a challenge. Scaling up to pilot-scale experiments is time-consuming, labour-intensive, and costly, with scaling losses being uncertain. This chapter discusses the application of computational fluid dynamics (CFD) in bio-electrochemical processes and the optimization of MFCs. It provides an overview of CFD-MFC modelling and simulation, discussing assumptions, models, and strategies to set a benchmark for future research. While few studies have integrated CFD modelling with MFC experimentation, the vast majority have focused on enhancing substrate mass transfer by enhancing the hydrodynamics and design of the anodic chamber. CFD's potential to advance MFC research is demonstrated by the congruence between these CFD modelling studies and actual experiments. Due to the novelty of CFD modelling for MFC systems, it is essential that future MFC-CFD research adheres to best practice guidelines to ensure accurate and reliable results. Thus, the application of CFD can provide an economically viable alternative to costly wet laboratory experimentation, thereby fostering further progress in MFC research.

2 Background

In recent years, the global energy demand has consistently risen, resulting in a pervasive energy crisis. Due to limited supplies and their detrimental environmental impact, including greenhouse gas emissions, reliance on fossil fuels, specifically oil, and gas, has become unsustainable. To combat this pressing problem, researchers have transferred their attention to the development of carbon-free, renewable, and alternative

S. E. Jujjavarapu et al., *Computational Fluid Dynamics Applications in Bio and Biomedical Processes*, https://doi.org/10.1007/978-981-99-7129-9_3

energy sources, investigating the viability of microbiology and biotechnology as potential solutions. Implementation of sustainable bioenergy is a promising strategy for addressing the difficulties posed by global warming (Ahmad & Zhang, 2020). In the light of this, there is a growing demand for new renewable energy technologies that can produce electricity without adding to net carbon dioxide emissions. In pursuance of a more sustainable and environmentally benign future, achieving zero net CO_2 emissions from energy production has become a highly desirable goal (Owusu & Asumadu-Sarkodie, 2016).

Organic waste, particularly biomass, is emerging as a promising and viable alternative energy source. Utilizing waste organics as biomass is both environmentally friendly and cost-effective (Alidrisi & Demirbas, 2016). Biomass energy derived from waste organics has a wide range of applications. It can be used to make ethanol, biodiesel, hydrogen fuel cells, and microbial fuel cells (MFCs), among other things (Kumar & Eswari, 2023). These advancements in biomass technology have enormous potential for sustainable and cleaner energy solutions. MFCs are bio-electrochemical devices that use electrogenic microorganisms to turn the chemical energy of organic materials into electricity. MFCs are made up of an anode, a cathode, and a proton exchange membrane. Organic substances are oxidized by microorganisms in the anode compartment, releasing electrons, protons, and carbon dioxide (CO_2). Electrons move to the cathode through an external circuit, where they combine with an electron acceptor to complete the electrochemical process. The proton exchange membrane enables protons to flow while maintaining an electrochemical gradient (Kumar & Eswari, 2023). The recent advancement of the MFC system, utilizing photosynthetic organisms known as photosynthetic MFCs, has transformed this system into a carbon–neutral system (Fig. 1) (Kumar & Jujjavarapu, 2023b). However, the research state of MFC development is still far from commercialization. A significant amount of fundamental research is still required to effectively apply MFC technology on an industrial scale. This study incorporates both experimental work and model simulations. As a result, optimizing and developing MFC technology requires interdisciplinary research combining expertise in electrochemistry, microbiology, biotechnology, and engineering (Boas et al., 2022).

CFD modelling and simulation are critical in the study and optimization of MFCs. CFD provides valuable insights into the system's behaviour by numerical modelling and simulation of fluid flow, species transport, and electrochemical reactions within the MFC. Understanding fluid dynamics improves mass transport to microorganisms and electron acceptors, ultimately improving MFC performance. Furthermore, CFD simulations predict current density and potential distribution, assisting in electrode design and maximizing power output. CFD modelling also aids in reactor design and scale-up, guiding the development of efficient and cost-effective MFC systems for industrial applications. Researchers can focus experimental efforts on key factors influencing MFC performance by conducting parametric studies. CFD modelling supplements experimental work, lowering the cost, and allowing researchers to optimize MFC technology and its potential for widespread use in a variety of industries

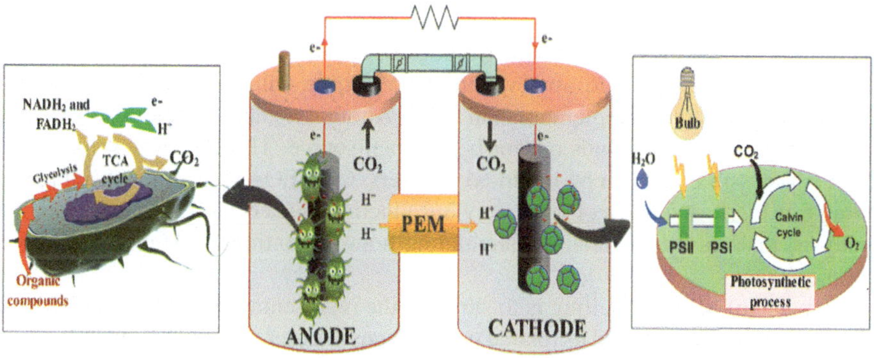

Fig. 1 Schematics and mechanisms of the photosynthetic microbial fuel cell (a zero carbon MFC system) (Kumar & Eswari Jujjavarappu, 2023)

and environmental applications (Kumar & Jujjavarapu, 2023a). The detailed understanding gained through simulations accelerates progress towards the development of efficient and sustainable MFCs.

Due to the limited number of studies integrating CFD in MFC experimentations, this chapter focuses on establishing a relationship between CFD and MFCs through a comprehensive analysis of previous studies and offering recommendations for future CFD modelling of MFCs. By implementing this approach, it presents a cost-effective substitute for conducting wet laboratory experiments in order to enhance and expand the efficiency of MFCs.

3 CFD Modelling Strategy for MFC Systems

There are various software options for performing CFD modelling and simulations. But among all CFD software, the widely utilized software packages are ANSYS CFX (cell-vertex) (Kim et al., 2014), ANSYS Fluent (cell-centred) (Rivera-Alvarez et al., 2020), and COMSOL Multiphysics (Kumar & Jujjavarapu, 2023a; Massaglia et al., 2017). These packages provide comprehensive CFD support, guiding users from preprocessing to post-processing stages. The primary difference between ANSYS and COMSOL CFD packages is their numerical methods for solving the relevant partial differential equations. The finite volume method (FVM) is used in ANSYS (Jeong & Seong, 2014), whereas the finite element method (FEM) is used in COMSOL (Ed Fontes, 2018). For CFD simulations, the FVM and FEM are referred to by their respective approaches to dividing the system into discrete elements. FEM divides the system into piecewise elements, whereas FVM divides it into control volumes. FVM is known for its conservatism, which ensures that the flux leaving one control volume is identical to the flux entering the next, making it ideally suited for solving conservation laws encountered in CFD (Dick, n.d.). Alternatively, FEM solvers, such as those found in COMSOL, are more suited for coupling CFD analysis with other simulations, such as electromagnetic and structural studies (Ed Fontes, 2018). It is important to note, however, that FEM requires more computing power and can be more difficult to implement than FVM. Another notable CFD software option is OpenFOAM, which stands out for being open-source. This means that its source code is accessible and can be modified by users. OpenFOAM has the advantage of being free to use and allows for the easy introduction of user-defined functions in the analysis, making it beneficial for simulating non-conventional systems like MFCs (Jasak, 2009).

The goal of MFC modelling is to simplify the representation of the complex bio-electrochemical system and its related mechanisms in order to improve understanding and clarity. However, due to the interdependence of microbial, electrochemical, and species transport factors, this task presents significant challenges. As a result, the development of models in the MFC community is frequently hindered by these intricate interactions. This section provides an overview of the various models used in MFC design and their fundamental assumptions. Additionally, it explores the practical application of these models in the design and operation of MFC processes.

3.1 Assumptions for MFC Modelling

Mathematical and numerical models are developed to capture the effect of various operating conditions and design parameters on the MFC system's performance. Due to the complexity of this bio-electrochemical system, it is necessary to make several assumptions in order to develop a simplified and linear representation of the intricate processes involved. Some of the common assumptions of CFD for MFC are listed below:

(i) The distribution of the electrolyte/substrate concentration across the electrode surface is uniform. The initial reaction is characterized by diffusion flux (J_i) at time zero (Kumar & Jujjavarapu, 2023a):

$$R_i = \frac{\Delta C_i}{\Delta t} + \Delta J_i + u * \Delta C_i \tag{1}$$

where *Ci* represents the initial electrolyte concentration, and *u* represents the velocity.

(ii) There is no flux in the boundary condition of the geometry of the electrode and the complete reaction occurs within the reference electrode, assuming an equilibrium state. (Kumar & Jujjavarapu, 2023a).

(iii) Throughout the operational period of the MFC, the anolyte and catholyte solutions are assumed to a homogenous (Luo et al., 2016).

(iv) In the cathodic chamber, a non-limiting reaction rate is taken into consideration (Luo et al., 2016).

(v) The electroneutrality within the biofilm is maintained by a non-limiting flow of anions and cations (Luo et al., 2016).
(vi) In many modelling approaches, they are developed based on a specific consideration, such as either electrochemical or biological factors, while keeping other parameters constant or under control. This allows researchers to focus on the effects of individual factors while simplifying the overall complexity of the model (Luo et al., 2016).
(vii) To minimize the impact of the environmental condition of the MFC, the operating conditions (such as pH and temperature) are assumed to be fully controlled (Radeef & Ismail, 2019).
(viii) Mathematical models are chosen based on the behaviour of microbial species in response to boundary conditions (Song et al., 2014).
(ix) The generation and transport of gases such as CO_2 within the anodic or cathodic chambers are neglected (Zeng et al., 2010).

3.2 Models for MFC Modelling

Researchers use a simplified form of ordinary differential equations and three-dimensional (3D) models incorporating fundamental equations such as the Nernst equation, the Butler–Volmer equation, Fick's law, Ohm's law, the Monod kinetic equation, individually or a combination of these to optimization of process and behaviour of MFCs. By combining these equations, it is possible to gain valuable insights into the electrochemical and biological processes that occur within MFCs, thereby facilitating the prediction of power generation capabilities and system performance (Ortiz-Martínez et al., 2015; Xia et al., 2018). The following subsection discussed the commonly employed models in MFCs:

(i) Monod kinetics equation: The linear Monod kinetics equation can effectively model the degradation of organic matter from the substrate via microbial metabolism. This equation describes the specific growth rate of microorganisms in response to substrate concentrations, allowing researchers to gain insight into the microbial activity involved in organic matter degradation. Using the linear Monod kinetics equation, one can better understand and predict the microbial processes within the system, leading to more accurate and comprehensive modelling of substrate degradation in a variety of applications, including MFCs (Deb et al., 2020). The linear Monod kinetics equation represents as follows:

$$\mu = \mu_{\max} + \frac{C_s}{K_s + C_s} \tag{2}$$

where $\mu_{\max}$ is a maximum specific growth rate, K_s is the substrate affinity constant, and C_s is substrate concentration.
(ii) Polarization model: Two fundamental equations, the Butler–Volmer equation (Eq. 3), and Ohm's law (Eq. 4), play critical roles in describing the electron

transfer mechanism and polarization behaviour in the modelling of MFCs. The Butler–Volmer equation characterizes the kinetics of electrode reactions, revealing the rates of anodic and cathodic processes at the electrodes (Dai et al., 2022; Zhao et al., 2016). Ohm's law, on the other hand, describes the relationship between electrical current, voltage, and resistance in an MFC system (Casula et al., 2021; Ou et al., 2017). Researchers can estimate the system's maximum energy recovery potential by incorporating these equations into the MFC model. The equations are expressed as follows:

$$J = J_0 \left\{ \exp\left[\frac{\alpha_a z F \eta}{RT} \right] - \exp\left[-\frac{\alpha_c z F \eta}{RT} \right] \right\} \tag{3}$$

where J is electrode current density (A/m^2), J_0 is exchange current density (A/m^2), T is absolute temperature (K), z is number of electrons involved in the electrode reaction, F is Faraday constant, R is universal gas constant, α_c is dimensionless cathodic charge transfer coefficient, α_a is dimensionless anodic charge transfer coefficient, and η is activation overpotential.

$$V = I * R \tag{4}$$

where V is Voltage, I is current, and R is a resistor.

(iii) Nernst equation: The Nernst equation plays a crucial role in the modelling of MFCs in estimating the redox potential under specific operating conditions. This equation establishes a relationship between the redox potential and several MFC variables, such as electron transfer kinetics and substrate species concentration. By incorporating these variables into the Nernst equation, researchers are able to gain insight into the electrochemical processes occurring within the MFC and predict the redox potential under the given conditions (Lim et al., 2021; Popat & Torres, 2016). The equations are typically represented in the following manner:

$$E = E_0 - \frac{RT}{(1-\alpha)nF} \ln Q \tag{5}$$

where E represents the reduction potential in volts (V) at the specific temperature and concentration of interest, E_0 represents the standard reduction potential in volts (V), R represents the universal constant, T represents the temperature in Kelvin (K), z representing the number of electrons transferred in the reaction, F represents the Faraday constant, and Q represents the quotient of the reaction.

(iv) Nernst–Planck equation: The Nernst–Planck equation is a fundamental mathematical model used to describe the ion transport mechanisms in MFCs. Understanding the transport of ions, such as protons and other charged species, across the membrane is essential for the efficient operation of the MFC and plays a crucial role in this regard. This equation is commonly used in MFC modelling to analyse the migration of ions across the membrane based on their concentration gradients and electric potential. Incorporating MFC variables associated with species and mass transport, such as ion concentration, charge density, and diffusion coefficients, the equation provides valuable insight into the electrochemical and transport processes occurring within the MFC system. Generally, the equation is expressed as follows:

$$F = \frac{-ZF}{RT} D_i \frac{\mathrm{dc}}{\mathrm{d}x} \tag{6}$$

where D denotes the diffusivity of chemical species, Z denotes the valence of ionic species, C denotes the concentration of the substrate. F denotes the Faraday constant, R denotes the universal constant, and T denotes the absolute temperature.

(v) Fick's Law: Fick's law is essential for characterizing analyte mass transport to the electrode surface. The analyte or substrate moves towards the electrode at a rate determined by the concentration difference between the bulk solution and the electrode. Fick's law is used in MFC modelling to understand and forecast the movement of different species, such as substrates or reactants, from the bulk solution to the electrode, hence impacting electrochemical reactions (Kumar & Jujjavarapu, 2023a). Researchers study diffusion-driven mass transfer processes and get insights into substrate availability and concentration profiles at the electrode–electrolyte interface by including Fick's law. The law states as follows:

$$\frac{\partial C_A}{\partial t} = D_A \frac{\partial^2 C_A}{\partial x^2} \tag{7}$$

The variables in the equation are as follows: x represents the distance from the electrode, t is the time, C_A indicates the reactant or substrate concentration, and D_A stands for the diffusion coefficient.

The restricted diffusion inside the confined Nernst diffusion area prevents the reactive analytes from diffusing into the bulk solution, preserving Nernstian equilibrium and allowing diffusion-controlled currents to be measured. In such cases, the use of Fick's Law for mass transfer diffusion simplifies calculations and leads to the computation of the peak current (P_i):

$$P_i = \left(2.69 * 10^5\right) n^{3/2} A C_A D_A^{1/2} V^{1/2} \tag{8}$$

where n representing the number of electrons gained during the reduction process, A denoting the working electrode's surface area (per unit), C_A indicating the molar concentration of species A, D_A the diffusion coefficient, and v, the scan rate (per unit).

(vi) Tafel equation: In the modelling of MFCs, the Tafel equation is combined with the Monod equations to describe the kinetics of anode and cathode reactions.

The Tafel equation is used to represent the electrode reaction kinetics, providing insights into electrocatalytic efficiency and the rate of electron transfer at the anode and cathode (Ortiz-Martínez et al., 2015). The equation is as follows:

$$E - E_{\text{eq}} = \frac{RT}{(1-\alpha)nF} \ln \frac{i}{i_0} \tag{9}$$

In the context of electrode reactions, the following variables are involved: E, the electrode potential; E_{eq}, the equilibrium potential; i, the current density; i_0, the exchange current density; T, the temperature; n, the number of electrons involved in the electrode reaction; F, the Faraday constant; R, the universal gas constant; and α, the charge transfer coefficient.

3.3 Model Simulation in CFD for MFC

CFD simulations provide a comprehensive approach to studying the multifaceted interactions in MFCs, revealing valuable information about hydrodynamics, mass transfer, and electrochemical processes. These simulations have important implications for optimizing MFC design, improving energy conversion efficiency, and advancing the long-term use of MFC technology in a variety of applications, including wastewater treatment and the production of clean energy. Various empirical and numerical CFD tools, such as COMSOL software, ANSYS Fluids, OpenFOAM, and others, are used for such computations. Figure 2 illustrates the overview of CFD modelling of MFCs. Additionally, the CFD modelling approaches for MFC are tabulated in Table 1. In this section, we summarize various CFD modellings and simulations for MFCs systems.

3.3.1 Reaction Kinetics Study

CFD modelling and simulation play a significant role in the study of reaction kinetics of MFC systems. The CFD studies enables researchers to model and simulate biochemical reactions, analyze mass transfer effects, and study electrochemical kinetics of MFC systems.

Fu et al. (2019) used a simplified 2D Butler–Volmer model in a CFD simulation to investigate the kinetic process of substrate degradation in a flow-through composite anode MFC. The goal was to gain a better understanding of how to improve the COD removal rate in such MFCs. The CFD simulation results showed that the composite anodes indeed exhibited a higher COD removal rate. Remarkably, the numerical COD concentration data obtained from the CFD simulation closely matched the experimental results (Fig. 3a), validating the model's efficacy. The results showed that combining wooden granular activated carbon and carbon cloth into a composite anode effectively reduced the COD concentration from 215 to 83 mg/L within 16 h. The

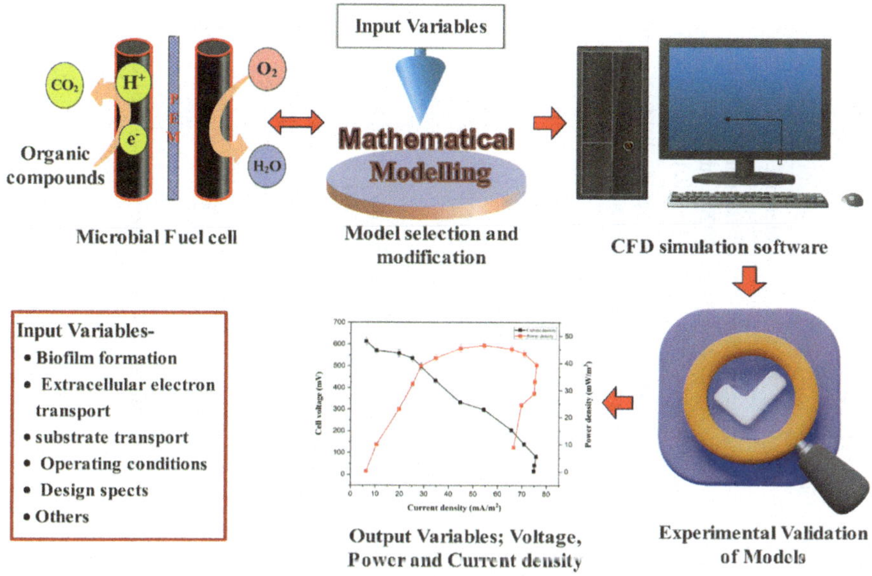

Fig. 2 Concept and outline of CFD modelling for MFC studies

contour plots of the dimensionless influent organic matter concentration are shown in Fig. 3b. Similarly, (Zhao et al., 2016) used a multiorder Butler–Volmer reaction model to calculate organic consumption and energy recovery in tubular MFCs. The hydrodynamics and species transport within the anodic compartment were studied using CFD. The simulation results revealed that the influent sodium acetate concentrations were 1.0, 0.5, and 0.3 g/L at anodic hydraulic retention times (HRTs) of 10 h. This demonstrates the system's effectiveness. Additionally, the maximum COD removal was found from 780 to 156 mg/L.

Moreover, (Sobieszuk et al., 2017) conducted flow simulations utilizing the CFD software ANSYS Fluent 17.0. CFD simulations were limited to the anode chamber as their computational domain. The authors made the assumption that the carbon fibre cloth covering the anode allowed for isothermal, laminar flow and that the fluid (wastewater) flux was not disturbed. The modelling results revealed that a Hydraulic Retention Time of 0.41 days resulted in the lowest value of Holdback (H), indicating optimal mixing conditions and the maximum COD reduction (55–60%) within the MFC (Fig. 4).

In a recent study published in 2021, (Farber et al., 2021) used a 3D Navier–Stokes equation in CFD ANSYS Fluent software to analyse fluid flow and the species balance equation for acetate in both the water and the biofilm (*Geobacter sulfurreducens*) in an MFC. The model included a representation of the acetate species mass fraction at the water–biofilm interface. The findings of this study revealed that the size of the biofilm's outer surface has a significant impact on the amount of electrical power generated by the biofilm. Notably, the outer surface of the biofilm, through which all

Table 1 Outlining the computational fluid dynamics (CFD) modelling inputs, processes, and outputs for an MFC

Aspect	Description	Input parameters	CFD process	Outputs	References
Geometry	Geometric representation of the MFC	Dimensions, shape, electrode arrangement	Mesh generation, boundary conditions	Mesh quality, flow field visualization	Bhaskaran and Collins, (2020), https://www.comsol.com/, (n.d.)
Operating conditions	Environmental factors affecting MFC performance	Temperature, pH, flow rate, substrate concentration	Flow simulation, species transport	Velocity distribution, concentration profiles	Kim et al. (2014), Kumar and Jujjavarapu (2023a)
Electrochemistry	Description of electrochemical reactions	Anode and cathode kinetics, EET mechanisms	Electrochemical reaction modelling	Current density distribution, overpotential analysis	Kumar and Jujjavarapu (2023a)
Porous media	Modelling of porous electrodes and biofilm growth	Porosity, tortuosity, biofilm thickness	Volume-averaged equations, biofilm modelling	Velocity through porous media, biofilm distribution	Li et al. (2022), Zhao et al. (2016)
Mass transfer	Movement of dissolved species in the MFC	Diffusion coefficients, mass transfer coefficients	Advection–diffusion modelling	Concentration profiles, diffusion fluxes	Kumar and Jujjavarapu (2023a)
Fluid flow	Movement of fluids within the MFC	Fluid viscosity, density, velocity boundary conditions	Navier–Stokes equations, turbulence modelling	Velocity field, pressure distribution	Kim et al. (2014), Massaglia et al. (2017), Zhao et al. (2016)
Electrode kinetics	Rate of electron transfer at the electrodes	Exchange current density, Butler–Volmer equation	Kinetic reaction modelling	Anodic and cathodic current densities, reaction rates	Ortiz-Martínez et al. (2015)
Heat transfer	Temperature distribution within the MFC	Thermal conductivity, heat generation rate	Energy equation, thermal boundary conditions	Temperature profiles, heat dissipation	Guzmán et al. (2018), Ortiz-Martínez et al. (2015)

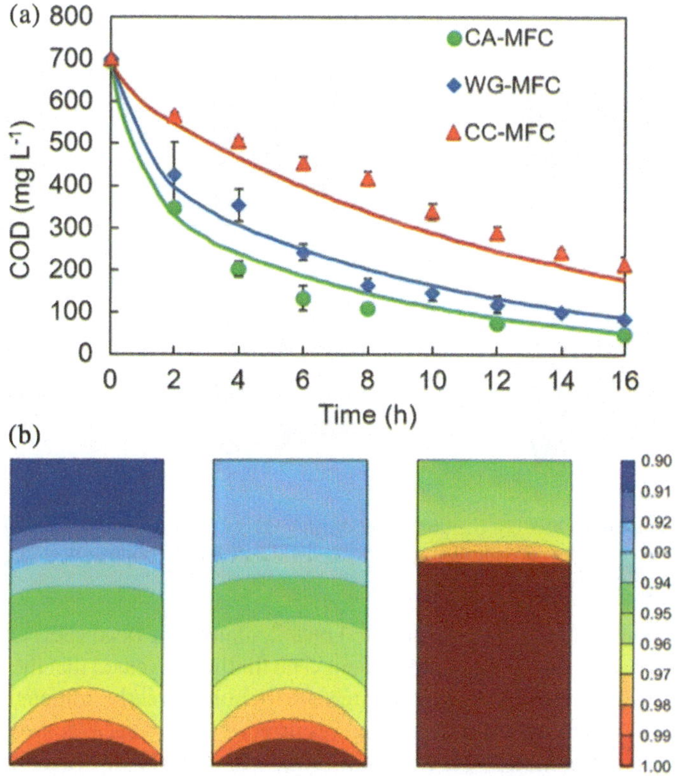

Fig. 3 This figure presents a comparison between experimental and numerical data of COD changes with time (**a**) for CA-MFC, WG-MFC, and CC-MFC. Additionally, it shows dimensionless contours of COD concentration for CA-MFC, WG-MFC, and CC-MFC (**b**) (Fu et al., 2019)

acetate is converted within the biofilm, appeared to be the critical factor influencing electrical power output.

3.3.2 Design of MFC

CFD and based models play a key role in the development of MFC systems. Some of the researchers used it to simulate fluid flow patterns, optimize internal structures, and evaluate various electrode geometries. It enables the investigation of various operating parameters and the identification of potential problems, thereby guiding design modifications. CFD speeds up design improvements by eliminating the need for costly experimental trials.

Kim et al. (2014) investigated various physicochemical phenomena, such as fluid flows, mass transfer, and chemical reactions, in twelve MFCs denoted as M1–M12. The MFCs were outfitted with distinct internal structures in order to generate more

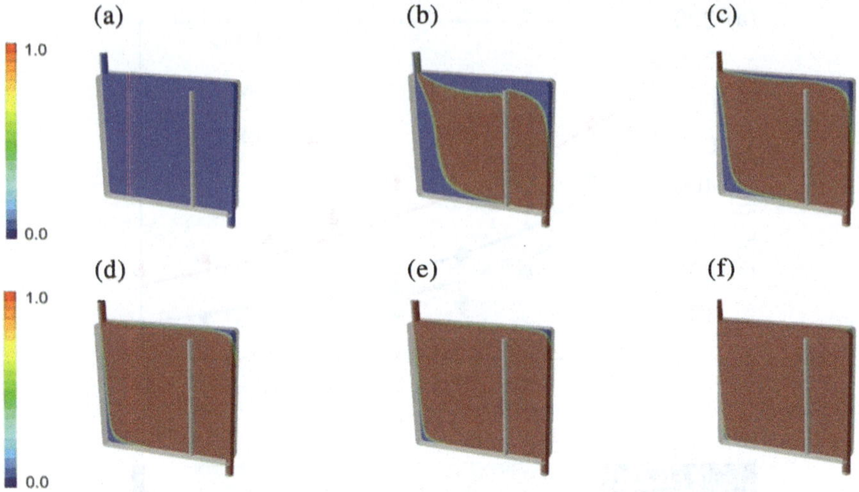

Fig. 4 Contours of the mixture fraction in anode of MFC, f (−), with time for $\tau = 0.41$ d; **a** $t = 0$, **b** $t = 0.5\ \tau$, **c** $t = 1.0\ \tau$, **d** $t = 3.0\ \tau$, **e** $t = 6.0\ \tau$, and **f** $t = 14.0\ \tau$ (Sobieszuk et al., 2017)

theoretical power. The study concentrated on calculating the dead (DS) and working (WS) spaces within the anode compartment. The findings revealed that the fluid flows varied significantly depending on the internal structure, directly influencing the WS where the bio-electrochemical reactions occurred. Notably, as WS size increased, power generation exhibited an exponential growth pattern. In order to maximize power generation, the study determined that MFC with 18 rectangular-type internal structures, denoted as M11, represented the best solution. In 2017, Massaglia et al. investigated the fluid dynamic distribution in two distinct geometries of air–cathode single chamber MFCs. These MFCs had a squared shape and a drop-like configuration, with an inner volume of 12.5 mL. The study's primary goal was to investigate their potential application as biosensors. The authors used COMSOL CFD software to simulate fluid dynamics using the Navier–Stokes equations. The simulations looked to determine the effective exposed area for each MFC architecture and to establish a relationship between this parameter and variations in device performance, particularly in terms of current densities. The study also looked at how the MFCs reacted to changes in sodium acetate concentration. The results showed that square SCMFCs with a non-symmetric inlet and outlet configuration (Sq2) had a higher percentage of exposed area than square SCMFCs with a symmetric inlet and outlet configuration (Sq1). Furthermore, the drop-like SCMFCs (Dr1) demonstrated the highest percentage of exposed area, with values of 96% and 94% for flow rates of 12.5 mL h^{-1} and 100 mL h^{-1}, respectively (Massaglia et al., 2017).

Sangeetha et al. (2021) investigated the effect of flow channel diameter on the performance of honeycomb MFC (HCMFCs) operating in recirculation batch mode in a similar manner. They estimated using CFD-ACE+ software (2015, (CFDRC) Research Corp., U.S.) and a flow turbulence model. Three different diameters, 0.4 cm,

0.7 cm, and 1 cm, were used in three different reactors referred to as HCMFC1, HCMFC 2, and HCMFC 3, respectively. They also included a control reactor without flow straighteners as a comparison. The HCMFC 2 model revealed that the flow inside the chamber had a medium-range velocity. The authors emphasized that this medium velocity flow was beneficial to reactor performance because it allowed for the uniform transfer of ions throughout the reactor. The experimental results were consistent with the computational simulations, with HCMFC 2 outperforming the others. It had the highest voltage generation of 0.55 V, the highest current density of 5300 mA/m^2, the highest power density of 430 mW/m^2, the highest organic content removal rate of 97.6%, the lowest internal resistance, and the thickest anode biofilm. Recently, (Kumar & Jujjavarapu, 2023a) used COMSOL 5.6 CFD software to design and simulate electroanalysis phenomena in double chamber MFCs. The primary focus of this study was on cyclic voltammetry and anolyte concentration distribution in seven different electrode geometries designs: hexagonal, square, pentagonal, circular, triangular, rectangular, and rhombus to enhance the power generation of MFCs (Fig. 5). This investigation used a model that included Fick's law and Ohm's law. CFD simulation results showed that the hexagonal design, which had a larger perimeter value and surface area of the working electrode, had a higher peak current (2422.75 mA) and a more uniform distribution of anolyte or substrate as compared to other designs. The experimental results corroborated the simulation results and demonstrated that the MFC setup with a hexagonal electrode produced the highest power density of 22.41 ± 0.32 mW/m^3 and the highest current density of 41.58 ± 0.35 mA/m^2.

4 Challenges and Future of CFD for MFC Modelling

4.1 Computational Complexity and Resource Requirements

The website (https://www.cfdsupport.com/hardware-for-cfd.html, n.d.) provides up-to-date configurations for CFD simulations. However, considering the complexity of CFD and limited software accessibility, we would like to offer some general tips and strategies that have proven helpful for researchers using CFD software in the modelling and simulation of biological systems including MFC. Remember that there are no minimum requirements for CFD simulations, and they can be run on a variety of devices such as personal computers, laptops, clusters, or even mobile devices. Although there are no hard and fast rules, it is recommended that you have a 64-bit system, 4 GB RAM memory, a 500 GB hard drive, and a 15-inch screen. It is critical to understand the scope of your simulation and its requirements. Consider mesh size, simulation time treatment (time method), and the complexity of the physical model. It is critical to balance these aspects with appropriate hardware for a successful CFD simulation (https://www.comsol.com/, n.d.).

Let's move on to hardware recommendations. To ensure satisfactory results for all parties involved, it is critical to strike a balance between hardware and simulation

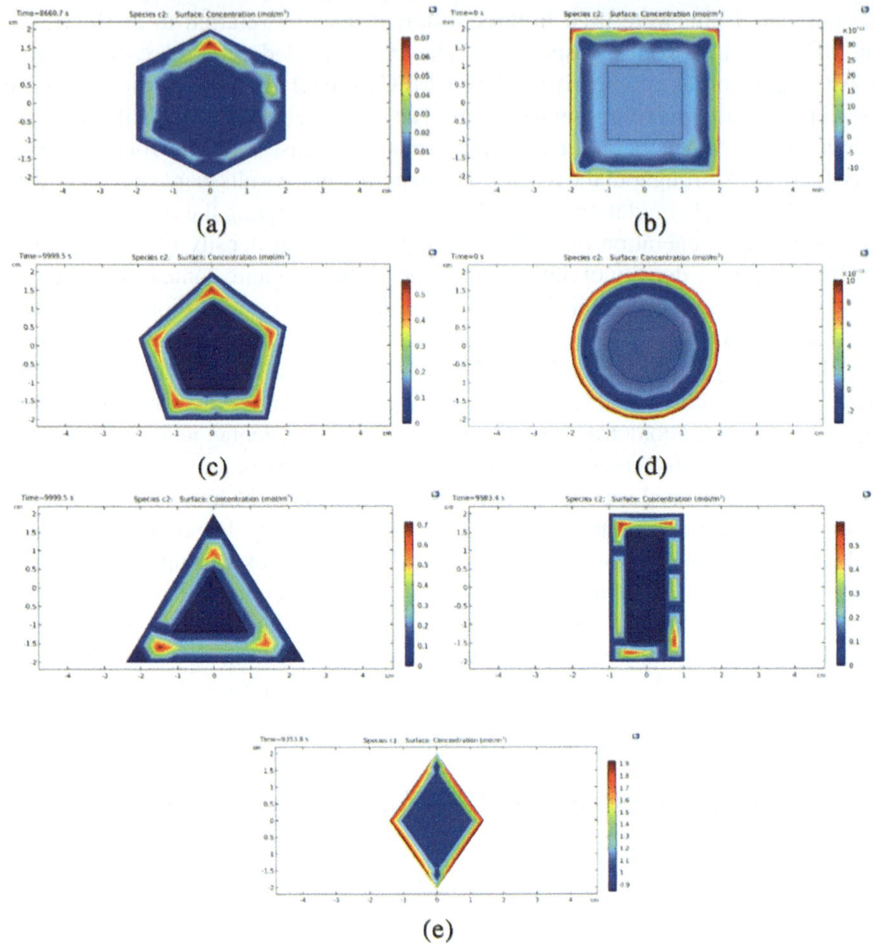

Fig. 5 Design and concentration profile of different geometry shapes of the working (anode) electrode **a** hexagonal, **b** square, **c** pentagonal, **d** circular, **e** triangular, **f** rectangular, and **g** rhombus (Kumar & Jujjavarapu, 2023a)

size. Waiting too long for results can have a negative impact on productivity and lead to missed opportunities. In these simulations, time is of the essence. The CPU processor is the most important piece of hardware. Its power and speed are critical in dealing with the workload. More CPU power is generally better, but finding a good power/cost ratio is critical for making efficient use of resources (Falcão, 1997; Mihalič et al., 2022). CFD simulations in this field can be computationally demanding due to the inherent complexity of bioprocess modelling. Many researchers use high-performance computing systems, parallel computing, and optimized solver algorithms to address this. These techniques aid in the acceleration of simulations and the effective management of resource requirements, allowing researchers to obtain results more quickly (D'Bastiani et al., 2023; Heroux, 2022).

4.2 CFD Model Validation

CFD model validation is an important aspect of CFD, as it ensures the accuracy and reliability of simulated results. The process entails comparing the model's predictions to experimental data or analytical solutions. CFD model validation has several limitations and challenges. These include the quality of experimental data, defining appropriate boundary conditions, complex turbulence modelling, sensitivity to mesh size and quality, and the necessity of model assumptions. Validation is also complicated by a lack of experimental data and scale effects. To overcome these constraints, caution is required, as well as sensitivity analyses and benchmarking against existing data. To improve accuracy and reliability, continuous improvements in turbulence models, meshing techniques, and boundary condition definitions are required. Using multiple validation methods can provide a more thorough evaluation of the model's performance (Alobaid et al., 2022; Lee et al., 2013).

CFD modelling approaches aim to capture the essence of electrochemical and biological complex reactions within a multispecies and multidimensional matrix, capturing the essence of these phenomena while minimizing plagiarism. Electrochemical models are based on well-established principles such as Monod kinetics, the Butler–Volmer equation, the Nernst equation, and the Tafel equation. These models provide a simplified representation of electrochemical half-cells, effectively balancing charge and mass within the system. However, the rate of redox reactions in MFCs is governed by interdependent anodic and cathodic parameters. Because the behaviour of one component affects the behaviour of the other, the correlation between these factors must be considered. However, existing CFD and mathematical models mostly focus on individual components or processes related to MFCs and fall short of integrating all of the complexities inherent in such a system.

To address this limitation and progress towards scaling-up applications, models that integrate each component harmoniously must be developed. Focusing solely on modelling either the anodic or cathodic component while holding the other constant fails to account for the intricate relationship between parameters and their impact on energy recovery in MFCs (Ortiz-Martínez et al., 2015; Sathe et al., 2021; Yi et al., 2020). One of the difficulties in achieving reliable results is the difficulty in replicating MFC performance, even under similar operating conditions. This is due to the system's robust and dynamic nature, which introduces errors during the validation of modelling outcomes (Kumar & Jujjavarapu, 2023a; Kumar et al., 2022).

Mathematical modelling is an important tool for understanding the processes involved in MFCs and identifying factors governing electricity generation for scaling-up strategies (Oh, 2010; Ortiz-Martínez et al., 2014). However, these models have previously overlooked the mechanisms underlying limiting processes in MFCs, necessitating further investigation through experimental validation. It is critical for aspiring researchers involved in upgrading existing models or developing new mathematical approaches for MFCs to address the limitations of current methods (Gadkari et al., 2018). Furthermore, when modelling scale-up systems, computational time must be considered and validated against experimental results. To achieve accurate simulation results, dedicated CFD modelling software such as COMSOL and a thorough understanding of the multidisciplinary system and boundary conditions are required (Kumar & Jujjavarapu, 2023a).

The future of CFD in the modelling of MFCs presents a promising outlook with numerous opportunities for advancement. CFD modelling will be critical in gaining a better understanding of the complex fluid flow and electrochemical processes that occur within MFCs. Researchers can gain more insight into the interactions between microbial communities, electrode materials, and the fluid environment as computational power and modelling techniques improve.

One of the most promising future prospects is the use of CFD simulations to optimize MFC designs and operational conditions with novel model creation and validation (Bhatti et al., 2020; Kamyar et al., 2012). Another promising direction is the incorporation of multiphysics phenomena into CFD models. A more accurate representation of real-world MFC behaviour will come from comprehensive modelling that takes into account microbial kinetics, mass transfer, heat transfer, and electrochemical reactions (Kone et al., 2017; Samstag et al., 2016a). For real-time monitoring and control of MFC systems, CFD models can be coupled with sensor data, allowing for optimized performance and rapid anomaly detection (Yi et al., 2020; Zhao et al., 2016). For increase the efficiency of MFCs, the CFD simulations can also help in designing high-performance and durable electrodes by nanoparticles by understanding material properties and interactions (Carpentieri et al., 2011; Lamon et al., 2019). Furthermore, the predictive capabilities of MFC performance can be improved by combining CFD modelling with artificial intelligence and machine learning algorithms, resulting in automated optimization processes and faster design iterations (Kumar et al., 2022).

5 Conclusion

The complexity of the MFC system can be effectively reduced by employing CFD and mathematical modelling, employing various electrochemical and microbial kinetic approaches. CFD can optimize the system and validate outputs against experimental results by incorporating multiple input variables, such as operating conditions and design variations. This study emphasized the importance of employing diverse modelling strategies in MFC research, based on their particular utility and purpose. This provides researchers with a clear direction for selecting the appropriate modelling approach in a systematic manner, with the goal of reducing computational time.

Future research endeavours should concentrate on developing novel mathematical models tailored to the complexities of MFC systems in order to advance towards scaling-up applications. Implementing these models in CFD simulations is essential for gaining an accurate understanding of the system's behaviour, design, and performance. This necessitates a thorough comprehension of the limitations associated with existing models in order to overcome potential obstacles and ensure the accuracy of future simulations. As research in this area continues to advance, the synergy between CFD and mathematical modelling will undoubtedly lead to ground-breaking innovations and accelerate the development of MFC and other biological system as a more efficient and viable energy source.

References

Abali, B. E. (2019). An accurate finite element method for the numerical solution of isothermal and incompressible flow of viscous fluid. *Fluids, 4*(1), 5. https://doi.org/10.3390/fluids4010005

Ahmad, T., & Zhang, D. (2020). A critical review of comparative global historical energy consumption and future demand: The story told so far. *Energy Reports, 6*, 1973–1991. https://doi.org/10.1016/j.egyr.2020.07.020

Ahmmed, M. S., Skerman, A., & Batstone, D. J. (2022). Predicting long-term solid accumulation in waste stabilisation lagoons through a combined CFD-process model approach. *Chemical Engineering Research and Design, 184*, 267–276. https://doi.org/10.1016/j.cherd.2022.06.012

Alidrisi, H., & Demirbas, A. (2016). Enhanced electricity generation using biomass materials. *Energy Sources, Part a: Recovery, Utilization, and Environmental Effects, 38*(10), 1419–1427. https://doi.org/10.1080/15567036.2014.948647

Alobaid, F., Almohammed, N., Massoudi Farid, M., May, J., Rößger, P., Richter, A., & Epple, B. (2022). Progress in CFD simulations of fluidized beds for chemical and energy process engineering. *Progress in Energy and Combustion Science, 91*, 100930. https://doi.org/10.1016/j.pecs.2021.100930

Aly, H. S., Eldrainy, Y. A., Saqr, K. M., Lazim, T. M., & Jaafar, M. N. M. (2010). A mathematical model for predicting spray atomization characteristics in an Eulerian-Eulerian framework. *International Communications in Heat and Mass Transfer, 37*(6), 618–623. https://doi.org/10.1016/j.icheatmasstransfer.2010.02.003

Amato, T., & Wicks, J. (2009). Dissolved air flotation and potential clarified water quality based on computational fluid dynamics modelling CFD with reactions-simulation of wastewater treatment plants view project. https://doi.org/10.13140/2.1.2605.2480

Arif, A. U. A., Sorour, M. T., & Aly, S. A. (2020). Cost analysis of activated sludge and membrane bioreactor WWTPs using CapdetWorks simulation program: Case study of Tikrit WWTP (middle Iraq). *Alexandria Engineering Journal, 59*(6), 4659–4667. https://doi.org/10.1016/j.aej.2020.08.023

Arumuga Perumal, D., & Dass, A. K. (2015). A Review on the development of lattice Boltzmann computation of macro fluid flows and heat transfer. *Alexandria Engineering Journal, 54*(4), 955–971. https://doi.org/10.1016/j.aej.2015.07.015

Azarpour, A., Rezaei, N., & Zendehboudi, S. (2021). Performance analysis and modeling of catalytic trickle-bed reactors: A comprehensive review. *Journal of Industrial and Engineering Chemistry, 103*, 1–41. https://doi.org/10.1016/j.jiec.2021.04.020

Bach, C., Yang, J., Larsson, H., Stocks, S. M., Gernaey, K. V., Albaek, M. O., & Krühne, U. (2017). Evaluation of mixing and mass transfer in a stirred pilot scale bioreactor utilizing CFD. *Chemical Engineering Science, 171*, 19–26. https://doi.org/10.1016/j.ces.2017.05.001

Balakin, B. V., Hoffmann, A. C., Kosinski, P., & Rhyne, L. D. (2010). Eulerian-Eulerian CFD model for the sedimentation of spherical particles in suspension with high particle concentrations.

Engineering Applications of Computational Fluid Mechanics, 4(1), 116–126. https://doi.org/10.1080/19942060.2010.11015303

Basavarajappa, M., & Miskovic, S. (2016). Investigation of gas dispersion characteristics in stirred tank and flotation cell using a corrected CFD-PBM quadrature-based moment method approach. *Minerals Engineering, 95*, 161–184. https://doi.org/10.1016/j.mineng.2016.06.026

Batstone, D. J., Puyol, D., Flores-Alsina, X., & Rodríguez, J. (2015). Mathematical modelling of anaerobic digestion processes: Applications and future needs. *Reviews in Environmental Science and Biotechnology, 14*(4), 595–613. https://doi.org/10.1007/s11157-015-9376-4

Behin, J., & Bahrami, S. (2012). Modeling an industrial dissolved air flotation tank used for separating oil from wastewater. *Chemical Engineering and Processing: Process Intensification, 59*, 1–8. https://doi.org/10.1016/j.cep.2012.05.004

Bhaskaran, R., & Collins, L. (2020). *Introduction to CFD Basics.*

Bhatti, M. M., Marin, M., Zeeshan, A., & Abdelsalam, S. I. (2020). Editorial: Recent trends in computational fluid dynamics. *Frontiers in Physics, 8*.https://doi.org/10.3389/fphy.2020.593111

Bhuyar, L. B., Thakre, S. B., & Ingole, N. W. (2009). Design characteristics of Curved Blade Aerator w.r.t. aeration efficiency and overall oxygen transfer coefficient and comparison with CFD modeling. *International Journal of Engineering, Science and Technology, 1*(1). www.ijest-ng.com

Boas, J. V., Oliveira, V. B., Simões, M., & Pinto, A. M. F. R. (2022). Review on microbial fuel cells applications, developments and costs. *Journal of Environmental Management, 307*, 114525. https://doi.org/10.1016/j.jenvman.2022.114525

Bouras, H., Haroun, Y., Fortunato Bodziony, F., Philippe, R., Fongarland, P., & Augier, F. (2022). Use of CFD for pressure drop, liquid saturation and wetting predictions in trickle bed reactors for different catalyst particle shapes. *Chemical Engineering Science, 249*, 117315. https://doi.org/10.1016/j.ces.2021.117315

Brannock, M., Wang, Y., & Leslie, G. (2010). Mixing characterisation of full-scale membrane bioreactors: CFD modelling with experimental validation. *Water Research, 44*(10), 3181–3191. https://doi.org/10.1016/j.watres.2010.02.029

Camp, T. R. (1942). Grit chamber design. *Source: Sewage Works Journal*, 14(2). https://about.jstor.org/terms

Cantarero Rivera, F. J., & Chen, J. (2022). Computational fluid dynamics modeling of cell cultures in bioreactors and its potential for cultivated meat production—A mini-review. *Future Foods, 6*, 100195. https://doi.org/10.1016/j.fufo.2022.100195

Carey, R. O., & Migliaccio, K. W. (2009). Contribution of wastewater treatment plant effluents to nutrient dynamics in aquatic systems. *Environmental Management, 44*(2), 205–217. https://doi.org/10.1007/s00267-009-9309-5

Carpentieri, M., Kumar, P., & Robins, A. (2011). An overview of experimental results and dispersion modelling of nanoparticles in the wake of moving vehicles. *Environmental Pollution, 159*(3), 685–693. https://doi.org/10.1016/j.envpol.2010.11.041

Casula, E., Molognoni, D., Borràs, E., & Mascia, M. (2021). 3D modelling of bioelectrochemical systems with brush anodes under fed-batch and flow conditions. *Journal of Power Sources, 487*, 229432. https://doi.org/10.1016/j.jpowsour.2020.229432

Chang, J., Zhang, K., Chen, H., Yang, Y., & Zhang, L. (2013). CFD modelling of the hydrodynamics and kinetic reactions in a fluidised-bed MTO reactor. *Chemical Engineering Research and Design, 91*(12), 2355–2368. https://doi.org/10.1016/J.CHERD.2013.04.023

Chen, B., Wei, G., Zhang, T., Wu, H., Wu, C., Chen, A., Zhang, H., Guan, X., Ren, Y., Feng, C., Qiu, G., & Wei, C. (2023). Flow drag force contributes high bio-treatment efficiency in a circulating fluidized bed reactor: Mechanism of selective separation of functional decayed sludge. *Chemical Engineering Journal, 454*, 140448. https://doi.org/10.1016/j.cej.2022.140448

Climent, J., Basiero, L., Martínez-Cuenca, R., Berlanga, J. G., Julián-López, B., & Chiva, S. (2018). Biological reactor retrofitting using CFD-ASM modelling. *Chemical Engineering Journal, 348*, 1–14. https://doi.org/10.1016/j.cej.2018.04.058

Cornelissen, J. T., Taghipour, F., Escudié, R., Ellis, N., & Grace, J. R. (2007). CFD modelling of a liquid–solid fluidized bed. *Chemical Engineering Science, 62*(22), 6334–6348. https://doi.org/10.1016/j.ces.2007.07.014

D'Bastiani, C., Kennedy, D., & Reynolds, A. (2023). CFD simulation of anaerobic granular sludge reactors: A review. *Water Research, 242*, 120220. https://doi.org/10.1016/j.watres.2023.120220

Dabiri, S., Kumar, P., & Rauch, W. (2023). Integrating biokinetics with computational fluid dynamics for energy performance analysis in anaerobic digestion. *Bioresource Technology, 373*, 128728. https://doi.org/10.1016/j.biortech.2023.128728

Dai, Z., Yu, R., Wu, Y., Zhu, G., Lu, X., & Zha, X. (2022). Modelling of self-sustainable microbial fuel cell type oil sensors based on restricted oxygen transfer and two-population competition. *Science of the Total Environment, 806*, 151333. https://doi.org/10.1016/j.scitotenv.2021.151333

Das, S., Bai, H., Wu, C., Kao, J. H., Barney, B., Kidd, M., & Kuettel, M. (2016). Improving the performance of industrial clarifiers using three-dimensional computational fluid dynamics. *Engineering Applications of Computational Fluid Mechanics, 10*(1), 130–144. https://doi.org/10.1080/19942060.2015.1121518

Deb, D., Patel, R., & Balas, V. E. (2020). A review of control-oriented bioelectrochemical mathematical models of microbial fuel cells. *Processes, 8*(5), 583. https://doi.org/10.3390/pr8050583

Dehnavi, M. A., Shahhosseini, S., Hashemabadi, S. H., & Ghafelebashi, S. M. (2010). CFD simulation of hydrodynamics and heat transfer in gas phase ethylene polymerization reactors. *International Communications in Heat and Mass Transfer, 37*(4), 437–442. https://doi.org/10.1016/j.icheatmasstransfer.2009.12.005

Dick, E. (n.d.). Introduction to finite volume methods in computational fluid dynamics. In *Computational Fluid Dynamics* (pp. 275–301). Springer Berlin Heidelberg. https://doi.org/10.1007/978-3-540-85056-4_11

Dick, E. (2009). Introduction to finite element methods in computational fluid dynamics. In *Computational Fluid Dynamics* (pp. 235–274). Springer Berlin Heidelberg. https://doi.org/10.1007/978-3-540-85056-4_10

Dixon, A. G., & Partopour, B. (2020). Computational fluid dynamics for fixed bed reactor design. *Annual Review of Chemical and Biomolecular Engineering, 11*(1), 109–130. https://doi.org/10.1146/annurev-chembioeng-092319-075328

Dong, Y., Sosna, B., Korup, O., Rosowski, F., & Horn, R. (2017). Investigation of radial heat transfer in a fixed-bed reactor: CFD simulations and profile measurements. *Chemical Engineering Journal, 317*, 204–214. https://doi.org/10.1016/j.cej.2017.02.063

Doumic, L. I., Soares, P. A., Ayude, M. A., Cassanello, M., Boaventura, R. A. R., & Vilar, V. J. P. (2015). Enhancement of a solar photo-Fenton reaction by using ferrioxalate complexes for the treatment of a synthetic cotton-textile dyeing wastewater. *Chemical Engineering Journal, 277*, 86–96. https://doi.org/10.1016/j.cej.2015.04.074

Dutta, A., Das, N., Sarkar, D., & Chakrabarti, S. (2019). Development and characterization of a continuous solar-collector-reactor for wastewater treatment by photo-Fenton process. *Solar Energy, 177*, 364–373. https://doi.org/10.1016/j.solener.2018.11.036

Ed Fontes. (2018, November 29). *FEM vs. FVM*. COMSOL.

Eshtiaghi, N., Markis, F., Yap, S. D., Baudez, J. C., & Slatter, P. (2013). Rheological characterisation of municipal sludge: A review. In *Water Research* (Vol. 47, Issue 15, pp. 5493–5510). Elsevier Ltd. https://doi.org/10.1016/j.watres.2013.07.001

Falcão, D. M. (1997). High performance computing in power system applications (pp. 1–23). https://doi.org/10.1007/3-540-62828-2_109

Farber, P., Gräbel, J., Kroppen, N., Pötschke, L., Roos, D., Rosenbaum, M., Stegschuster, G., & Ueberholz, P. (2021). Electricity generation in a microbial fuel cell with textile carbon fibre anodes. *Computers & Mathematics with Applications, 83*, 4–23. https://doi.org/10.1016/j.camwa.2019.11.019

Farzan, P., Mistry, B., & Ierapetritou, M. G. (2017). Review of the important challenges and opportunities related to modeling of mammalian cell bioreactors. *AIChE Journal, 63*(2), 398–408. https://doi.org/10.1002/aic.15442

Feickert Fenske, C., Strübing, D., & Koch, K. (2023). Biological methanation in trickle bed reactors—A critical review. *Bioresource Technology, 385*, 129383. https://doi.org/10.1016/j.biortech.2023.129383

Fu, B., Xu, T., Guo, X., Liang, P., Huang, X., & Zhang, X. (2019). Optimization and simulation of a carbon-based flow-through composite anode configuration to enhance power generation and improve effluent quality simultaneously for microbial fuel cells. *Journal of Cleaner Production, 229*, 542–551. https://doi.org/10.1016/j.jclepro.2019.04.308

GM, K., & Buwa, V. V. (2018). Particle-resolved simulations of methane steam reforming in multilayered packed beds. *AIChE Journal, 64*(11), 4162–4176.https://doi.org/10.1002/aic.16386

Gadkari, S., Gu, S., & Sadhukhan, J. (2018). Towards automated design of bioelectrochemical systems: A comprehensive review of mathematical models. *Chemical Engineering Journal, 343*, 303–316. https://doi.org/10.1016/j.cej.2018.03.005

Gao, H., & Stenstrom, M. K. (2020). Development and applications in computational fluid dynamics modeling for secondary settling tanks over the last three decades: A review. In *Water Environment Research* (Vol. 92, Issue 6, pp. 796–820). John Wiley and Sons Inc. https://doi.org/10.1002/wer.1279

Glessmer, M., & Janßen, C. (2017). Using an interactive Lattice Boltzmann solver in fluid mechanics instruction. *Computation, 5*(4), 35. https://doi.org/10.3390/computation5030035

Gómez-Ríos, D., Junne, S., Neubauer, P., Ochoa, S., Ríos-Estepa, R., & Ramírez-Malule, H. (2019). Characterization of the Metabolic Response of Streptomyces clavuligerus to Shear Stress in Stirred Tanks and Single-Use 2D Rocking Motion Bioreactors for Clavulanic Acid Production. *Antibiotics, 8*(4), 168. https://doi.org/10.3390/antibiotics8040168

Goswami, L., Vinoth Kumar, R., Borah, S. N., Arul Manikandan, N., Pakshirajan, K., & Pugazhenthi, G. (2018). Membrane bioreactor and integrated membrane bioreactor systems for micropollutant removal from wastewater: A review. *Journal of Water Process Engineering, 26*, 314–328. https://doi.org/10.1016/j.jwpe.2018.10.024

Gu, Y., Li, Y., Yuan, F., & Yang, Q. (2023). Optimization and control strategies of aeration in WWTPs: A review. *Journal of Cleaner Production*, 138008.https://doi.org/10.1016/j.jclepro.2023.138008

Guzmán, C., Carrera, J., Durán, H., Berumen, J., Ortiz, A., Guirette, O., Arroyo, A., Brizuela, J., Gómez, F., Blanco, A., Azcaray, H., & Hernández, M. (2018). Implementation of virtual sensors for monitoring temperature in greenhouses using CFD and control. *Sensors, 19*(1), 60. https://doi.org/10.3390/s19010060

Hassanzadeh, A., Firouzi, M., Albijanic, B., & Celik, M. S. (2018). A review on determination of particle–bubble encounter using analytical, experimental and numerical methods. *Minerals Engineering, 122*, 296–311. https://doi.org/10.1016/j.mineng.2018.04.014

He, W., Li, Z. K., Tang, A. R., Liu, A., & He, B. G. (2013). Simulation study of three-phase flow field based on microbubble flotation. *Applied Mechanics and Materials, 477–478*, 226–231. https://doi.org/10.4028/www.scientific.net/AMM.477-478.226

Heroux, M. A. (2022). Research software science: Expanding the impact of research software engineering. *Computing in Science & Engineering, 24*(6), 22–27. https://doi.org/10.1109/MCSE.2023.3260475

Hickman, D. A., Degenstein, J. C., & Ribeiro, F. H. (2016). Fundamental principles of laboratory fixed bed reactor design. *Current Opinion in Chemical Engineering, 13*, 1–9. https://doi.org/10.1016/j.coche.2016.07.002

Hoiberg, B., & Shah, M. T. (2021). CFD study of multiphase flow in aerated grit tank. *Journal of Water Process Engineering, 39*https://doi.org/10.1016/j.jwpe.2020.101698

Hosseini, S. H., Ahmadi, G., & Olazar, M. (2013). CFD simulation of cylindrical spouted beds by the kinetic theory of granular flow. *Powder Technology, 246*, 303–316. https://doi.org/10.1016/j.powtec.2013.05.016

Hosseini, S. H., Rahimi, R., Zivdar, M., & Samimi, A. (2009). The effect of ring baffles on the hydrodynamics of a gas—Solid bubbling fluidized bed using computational fluid dynamics. *Proceedings of the Institution of Mechanical Engineers, Part c: Journal of Mechanical Engineering Science, 223*(10), 2281–2289. https://doi.org/10.1243/09544062JMES1575

Hreiz, R., Potier, O., Wicks, J., & Commenge, J. M. (2019). CFD Investigation of the effects of bubble aerator layouts on hydrodynamics of an activated sludge channel reactor. *Environmental Technology (united Kingdom), 40*(20), 2657–2670. https://doi.org/10.1080/09593330.2018.1448001

https://www.cfdsupport.com/hardware-for-cfd.html. (n.d.). *Hardware for CFD simulations.*

https://www.comsol.com/. (n.d.). COMSOL Multiphysics® Simulation Software Understand, Predict, and Optimize Real-World Designs, Devices, and Processes with Simulation.

Hyre, M. (2012). *Evaluation of three full-scale grit removal processes using CFD modeling*. https://www.researchgate.net/publication/265786513

Itoh, N., & Mimura, K. (2013). Computational fluid dynamics (CFD) analysis of membrane reactors: Simulation of single-and multi-tube palladium membrane reactors for hydrogen recovery from cyclohexane. In *Handbook of Membrane Reactors* (Vol. 1, pp. 464–495). Elsevier Inc. https://doi.org/10.1533/9780857097330.3.464

Jafarinejad, S. (2017). Cost estimation and economical evaluation of three configurations of activated sludge process for a wastewater treatment plant (WWTP) using simulation. *Applied Water Science, 7*(5), 2513–2521. https://doi.org/10.1007/S13201-016-0446-8

Jalilnejad, E., Jabbari, B., & Ghasemzadeh, K. (2022). Application of computational fluid dynamics technique in membrane bioreactor systems. In *Current Trends and Future Developments on (Bio-) Membranes* (pp. 345–375). Elsevier. https://doi.org/10.1016/B978-0-12-822294-2.00007-2

Jamshidian, R., Scully, J., & Van den Akker, H. E. A. (2023). Two-fluid simulations of an aerated lab-scale bioreactor. *Chemical Engineering Research and Design, 196*, 254–275. https://doi.org/10.1016/j.cherd.2023.06.042

Jasak, H. (2009). OpenFOAM: Open source CFD in research and industry. *International Journal of Naval Architecture and Ocean Engineering, 1*(2), 89–94. https://doi.org/10.2478/IJNAOE-2013-0011

Jasak, H., & Uroić, T. (2020). *Practical computational fluid dynamics with the finite volume method* (pp. 103–161). CISM International Centre for Mechanical Sciences. https://doi.org/10.1007/978-3-030-37518-8_4

Jeong, W., & Seong, J. (2014). Comparison of effects on technical variances of computational fluid dynamics (CFD) software based on finite element and finite volume methods. *International Journal of Mechanical Sciences, 78*, 19–26. https://doi.org/10.1016/j.ijmecsci.2013.10.017

Jin, Y., Liu, C. L., Song, X. F., & Yu, J. G. (2019). Computational fluid dynamics simulation as a tool for optimizing the hydrodynamic performance of membrane bioreactors. *RSC Advances, 9*(55), 32034–32046. https://doi.org/10.1039/c9ra06706j

John, D. (n.d.). *Improved design of settling tanks using an extended drift flux model*. http://research-information.bristol.ac.uk

Jurtz, N., Kraume, M., & Wehinger, G. D. (2019). Advances in fixed-bed reactor modeling using particle-resolved computational fluid dynamics (CFD). *Reviews in Chemical Engineering, 35*(2), 139–190. https://doi.org/10.1515/revce-2017-0059

Kamyar, A., Saidur, R., & Hasanuzzaman, M. (2012). Application of computational fluid dynamics (CFD) for nanofluids. *International Journal of Heat and Mass Transfer, 55*(15–16), 4104–4115. https://doi.org/10.1016/j.ijheatmasstransfer.2012.03.052

Karn, A. L., Pandya, S., Mehbodniya, A., Arslan, F., Sharma, D. K., Phasinam, K., Aftab, M. N., Rajan, R., Bommisetti, R. K., & Sengan, S. (2023). An integrated approach for sustainable development of wastewater treatment and management system using IoT in smart cities. *Soft Computing, 27*(8), 5159–5175. https://doi.org/10.1007/s00500-021-06244-9

Karpinska, A. M., & Bridgeman, J. (2016a). CFD-aided modelling of activated sludge systems—A critical review. *Water Research, 88*, 861–879. https://doi.org/10.1016/j.watres.2015.11.008

Karpinska, A. M., & Bridgeman, J. (2016b). CFD-aided modelling of activated sludge systems—A critical review. In *Water Research* (Vol. 88, pp. 861–879). Elsevier Ltd. https://doi.org/10.1016/j.watres.2015.11.008

Kim, J., Kim, H., Kim, B., & Yu, J. (2014). Computational fluid dynamics analysis in microbial fuel cells with different anode configurations. *Water Science and Technology, 69*(7), 1447–1452. https://doi.org/10.2166/wst.2014.041

Koerich, D. M., Lopes, G. C., & Rosa, L. M. (2018). Investigation of phases interactions and modification of drag models for liquid-solid fluidized bed tapered bioreactors. *Powder Technology, 339*, 90–101. https://doi.org/10.1016/j.powtec.2018.07.102

Kone, J.-P., Zhang, X., Yan, Y., Hu, G., & Ahmadi, G. (2017). Three-dimensional multiphase flow computational fluid dynamics models for proton exchange membrane fuel cell: A theoretical development. *The Journal of Computational Multiphase Flows, 9*(1), 3–25. https://doi.org/10.1177/1757482X17692341

Korzani, M. G., Galindo Torres, S., Scheuermann, A., & Williams, D. J. (2016). Smoothed particle hydrodynamics into the fluid dynamics of classical problems. *Applied Mechanics and Materials, 846*, 73–78. https://doi.org/10.4028/www.scientific.net/AMM.846.73

Kostoglou, M., Karapantsios, T. D., & Matis, K. A. (2007). CFD model for the design of large scale flotation tanks for water and wastewater treatment. *Industrial and Engineering Chemistry Research, 46*(20), 6590–6599. https://doi.org/10.1021/ie0703989

Kumar, T., & Eswari, J. S. (2023). Review and perspectives of emerging green technology for the sequestration of carbon dioxide into value-added products: An intensifying development. *Energy & Fuels, 37*(5), 3570–3589. https://doi.org/10.1021/acs.energyfuels.2c04122

Kumar, T., & Eswari Jujjavarappu, S. (2023). A critical review on an advanced bio-electrochemical system for carbon dioxide sequestration and wastewater treatment. *Total Environment Research Themes, 5*, 100023. https://doi.org/10.1016/j.totert.2022.100023

Kumar, T., & Jujjavarapu, S. E. (2023a). Advances in anode configurations for a microbial fuel cell via a computational fluid dynamics electrochemistry and its experimental validation. *Journal of Chemical Technology & Biotechnology*. https://doi.org/10.1002/jctb.7397

Kumar, T., & Jujjavarapu, S. E. (2023b). Carbon dioxide sequestration and wastewater treatment via an innovative self-sustaining algal microbial fuel cell. *Journal of Cleaner Production, 415*, 137836. https://doi.org/10.1016/j.jclepro.2023.137836

Kumar, T., Naik, S., & Jujjavarappu, S. E. (2021). A critical review on early-warning electrochemical system on microbial fuel cell-based biosensor for on-site water quality monitoring. *Chemosphere*, 133098.https://doi.org/10.1016/j.chemosphere.2021.133098

Kumar, T., Naik, S., & Jujjavarappu, S. E. (2022). A critical review on early-warning electrochemical system on microbial fuel cell-based biosensor for on-site water quality monitoring. *Chemosphere, 291*, 133098. https://doi.org/10.1016/j.chemosphere.2021.133098

Kwon, S. B., Park, N. S., Lee, S. J., Ahn, H. W., & Wang, C. K. (2006). Examining the effect of length/width ratio on the hydro-dynamic behaviour in a DAF system using CFD and ADV techniques. *Water Science and Technology, 53*(7), 141–149. https://doi.org/10.2166/wst.2006.218

Lamon, L., Asturiol, D., Vilchez, A., Cabellos, J., Damásio, J., Janer, G., Richarz, A., & Worth, A. (2019). Physiologically based mathematical models of nanomaterials for regulatory toxicology: A review. *Computational Toxicology, 9*, 133–142. https://doi.org/10.1016/j.comtox.2018.10.002

le Moullec, Y., Gentric, C., Potier, O., & Leclerc, J. P. (2010). CFD simulation of the hydrodynamics and reactions in an activated sludge channel reactor of wastewater treatment. *Chemical Engineering Science, 65*(1), 492–498. https://doi.org/10.1016/j.ces.2009.03.021

Lee, I.-B., Bitog, J. P. P., Hong, S.-W., Seo, I.-H., Kwon, K.-S., Bartzanas, T., & Kacira, M. (2013). The past, present and future of CFD for agro-environmental applications. *Computers and Electronics in Agriculture, 93*, 168–183. https://doi.org/10.1016/j.compag.2012.09.006

Lee, K. H., Kim, H., KuK, J. W., Chung, J. D., Park, S., & Kwon, E. E. (2020). Micro-bubble flow simulation of dissolved air flotation process for water treatment using computational fluid dynamics technique. *Environmental Pollution, 256*.https://doi.org/10.1016/j.envpol.2019.01.011

Lee, J. L., & Lim, E. W. C. (2017). Comparisons of Eulerian-Eulerian and CFD-DEM simulations of mixing behaviors in bubbling fluidized beds. *Powder Technology, 318*, 193–205. https://doi.org/10.1016/j.powtec.2017.05.050

Lettieri, P., & Mazzei, L. (2009). Challenges and issues on the CFD modeling of fluidized beds: A review. *The Journal of Computational Multiphase Flows, 1*(2), 83–131. https://doi.org/10.1260/175748209789563937

Li, L., Wang, K., Zhao, Q., Gao, Q., Zhou, H., Jiang, J., & Mei, W. (2022). A critical review of experimental and CFD techniques to characterize the mixing performance of anaerobic digesters for biogas production. *Reviews in Environmental Science and Bio/technology, 21*(3), 665–689. https://doi.org/10.1007/s11157-022-09626-z

Li, W., Li, M., Bolton, J. R., & Qiang, Z. (2016). Configuration optimization of UV reactors for water disinfection with computational fluid dynamics: Feasibility of using particle minimum UV dose as a performance indicator. *Chemical Engineering Journal, 306*, 1–8. https://doi.org/10.1016/j.cej.2016.07.042

Li, W., Li, M., Wen, D., & Qiang, Z. (2018a). Development of economical-running strategy for multi-lamp UV disinfection reactors in secondary water supply systems with computational fluid dynamics simulations. *Chemical Engineering Journal, 343*, 317–323. https://doi.org/10.1016/j.cej.2018.03.017

Li, X., Scott, K., Kelly, W. J., & Huang, Z. (2018b). Development of a computational fluid dynamics model for scaling-up Ambr bioreactors. *Biotechnology and Bioprocess Engineering, 23*(6), 710–725. https://doi.org/10.1007/s12257-018-0063-5

Lim, S. S., Fontmorin, J.-M., Pham, H. T., Milner, E., Abdul, P. M., Scott, K., Head, I., & Yu, E. H. (2021). Zinc removal and recovery from industrial wastewater with a microbial fuel cell: Experimental investigation and theoretical prediction. *Science of the Total Environment, 776*, 145934. https://doi.org/10.1016/j.scitotenv.2021.145934

Lin, J.-R., Cao, J., Zhang, J.-P., van Treeck, C., & Frisch, J. (2019). Visualization of indoor thermal environment on mobile devices based on augmented reality and computational fluid dynamics. *Automation in Construction, 103*, 26–40. https://doi.org/10.1016/j.autcon.2019.02.007

Liu, X., Wang, Y., Shi, Y., Li, Q., Dai, P., Guan, J., Waite, T. D., & Leslie, G. (2019). CFD modelling of uneven flows behaviour in flat-sheet membrane bioreactors: From bubble generation to shear stress distribution. *Journal of Membrane Science, 570–571*, 146–155. https://doi.org/10.1016/j.memsci.2018.10.040

Liu, X., Wang, Y., Waite, T. D., & Leslie, G. (2015). Numerical simulation of bubble induced shear in membrane bioreactors: Effects of mixed liquor rheology and membrane configuration. *Water Research, 75*, 131–145. https://doi.org/10.1016/j.watres.2015.02.009

Llano-Serna, C. E., Coral-Portillo, J. A., Fontalvo, J., & Prado-Rubio, O. A. (2019). CFD assisted analysis and design of hydraulic flocculators. *Revista Mexicana de Ingeniera Quimica, 18*(3), 995–1015. https://doi.org/10.24275/UAM/IZT/DCBI/REVMEXINGQUIM/2019V18N3/LLANO

Logtenberg, S. A., & Dixon, A. G. (1998). Computational fluid dynamics studies of fixed bed heat transfer. *Chemical Engineering and Processing: Process Intensification, 37*(1), 7–21. https://doi.org/10.1016/S0255-2701(97)00032-9

Lopes, R. J. G., & Quinta-Ferreira, R. M. (2007). Trickle-bed CFD studies in the catalytic wet oxidation of phenolic acids. *Chemical Engineering Science, 62*(24), 7045–7052. https://doi.org/10.1016/j.ces.2007.08.085

Lopes, R. J. G., Silva, A. M. T., & Quinta-Ferreira, R. M. (2007). Kinetic modeling and trickle-bed CFD studies in the catalytic wet oxidation of vanillic acid. *Industrial & Engineering Chemistry Research, 46*(25), 8380–8387. https://doi.org/10.1021/ie070009a

López-Jiménez, P. A., Escudero-González, J., Montoya Martínez, T., Fajardo Montañana, V., & Gualtieri, C. (2015). Application of CFD methods to an anaerobic digester: The case of Ontinyent WWTP, Valencia, Spain. *Journal of Water Process Engineering, 7*, 131–140. https://doi.org/10.1016/j.jwpe.2015.05.006

Luo, S., Sun, H., Ping, Q., Jin, R., & He, Z. (2016). A review of modeling bioelectrochemical systems: engineering and statistical aspects. *Energies, 9*(2), 111. https://doi.org/10.3390/en9020111

Macfarlan, L. H., Phan, M. T., & Bruce Eldridge, R. (2021). Structured packing geometry study for liquid-phase mass transfer and hydrodynamic performance using CFD. *Chemical Engineering Science*, 117353https://doi.org/10.1016/J.CES.2021.117353

Makatsa, T. J., Baloyi, J., Ntho, T., & Masuku, C. M. (2021a). Catalytic wet air oxidation of phenol: Review of the reaction mechanism, kinetics, and CFD modeling. *Critical Reviews in Environmental Science and Technology, 51*(17), 1891–1923. https://doi.org/10.1080/10643389.2020.1771886

Makatsa, T. J., Baloyi, S. J., & Masuku, C. M. (2021). *Computational fluid dynamics modelling of phenol oxidation in a trickle-bed reactor using 3D eulerian model* (pp. 825–831). https://doi.org/10.1016/B978-0-323-88506-5.50129-7

Malang, J., Kumar, P., & Saptoro, A. (2015). Computational fluid dynamics-based hydrodynamics studies in packed bed columns: Current status and future directions. *International Journal of Chemical Reactor Engineering, 13*(3), 289–303. https://doi.org/10.1515/ijcre-2014-0121

Malato, S., Fernández-Ibáñez, P., Maldonado, M. I., Blanco, J., & Gernjak, W. (2009). Decontamination and disinfection of water by solar photocatalysis: Recent overview and trends. In *Catalysis Today* (Vol. 147, Issue 1, pp. 1–59). https://doi.org/10.1016/j.cattod.2009.06.018

Mandenius, C.-F. (2016). Challenges for bioreactor design and operation. In *Bioreactors* (pp. 1–34). Wiley-VCH Verlag GmbH & Co. KGaA. https://doi.org/10.1002/9783527683369.ch1

Massaglia, G., Gerosa, M., Agostino, V., Cingolani, A., Sacco, A., Saracco, G., Margaria, V., & Quaglio, M. (2017). Fluid dynamic modeling for microbial fuel cell based biosensor optimization. *Fuel Cells, 17*(5), 627–634. https://doi.org/10.1002/fuce.201700026

Matko, T., Chew, J., Wenk, J., Chang, J., & Hofman, J. (2021). Computational fluid dynamics simulation of two-phase flow and dissolved oxygen in a wastewater treatment oxidation ditch. *Process Safety and Environmental Protection, 145*, 340–353. https://doi.org/10.1016/j.psep.2020.08.017

Mehl, M., Neckel, T., & Neumann, Ph. (2011). Navier-Stokes and Lattice-Boltzmann on octree-like grids in the Peano framework. *International Journal for Numerical Methods in Fluids, 65*(1–3), 67–86.https://doi.org/10.1002/fld.2469

Meroney, R. N., & Sheker, R. E. (2015). Removing grit during wastewater treatment: CFD analysis of HDVS performance. *Water Environment Research, 88*(5), 438–448. https://doi.org/10.2175/106143015x14212658614478

Mihalič, F., Truntič, M., & Hren, A. (2022). Hardware-in-the-loop simulations: A historical overview of engineering challenges. *Electronics, 11*(15), 2462. https://doi.org/10.3390/electronics11152462

Minin, O. (Ed.). (2011). *Computational fluid dynamics technologies and applications*. InTech. https://doi.org/10.5772/686

Morchain, J., Gabelle, J.-C., & Cockx, A. (2013). Coupling of biokinetic and population balance models to account for biological heterogeneity in bioreactors. *AIChE Journal, 59*(2), 369–379. https://doi.org/10.1002/aic.13820

Moreira, F. C., Boaventura, R. A. R., Brillas, E., & Vilar, V. J. P. (2017). Electrochemical advanced oxidation processes: A review on their application to synthetic and real wastewaters. In *Applied Catalysis B: Environmental* (Vol. 202, pp. 217–261). Elsevier B.V. https://doi.org/10.1016/j.apcatb.2016.08.037

Moustakas, K., Rehan, M., Loizidou, M., Nizami, A. S., & Naqvi, M. (2020). Energy and resource recovery through integrated sustainable waste management. *Applied Energy, 261*, 114372. https://doi.org/10.1016/j.apenergy.2019.114372

Mrope, H. A., Chande Jande, Y. A., & Kivevele, T. T. (2021). A Review on computational fluid dynamics applications in the design and optimization of crossflow hydro turbines. *Journal of Renewable Energy, 2021*, 1–13. https://doi.org/10.1155/2021/5570848

Muhammad, N. (2021). Finite volume method for simulation of flowing fluid via OpenFOAM. *The European Physical Journal plus, 136*(10), 1010. https://doi.org/10.1140/epjp/s13360-021-01983-y

Nadal-Rey, G., McClure, D. D., Kavanagh, J. M., Cornelissen, S., Fletcher, D. F., & Gernaey, K. V. (2021). Understanding gradients in industrial bioreactors. *Biotechnology Advances, 46*, 107660. https://doi.org/10.1016/j.biotechadv.2020.107660

Naik, S., & Jujjavarappu, S. E. (2020). Simultaneous bioelectricity generation from cost-effective MFC and water treatment using various wastewater samples. *Environmental Science and Pollution Research, 27*(22), 27383–27393. https://doi.org/10.1007/s11356-019-06221-8

Nakatsuka, N., Kishita, Y., Kurafuchi, T., & Akamatsu, F. (2020). Integrating wastewater treatment and incineration plants for energy-efficient urban biomass utilization: A life cycle analysis. *Journal of Cleaner Production, 243*, 118448. https://doi.org/10.1016/J.JCLEPRO.2019.118448

Norouzi-Firouz, H., Sarrafzadeh, M., Zarghami, R., & Moshiri-Tabrizi, I. (2022). A coupled hydrodynamic-biokinetic simulation of three-phase flow in an oxidation ditch using CFD. *The Canadian Journal of Chemical Engineering, 100*(2), 223–236. https://doi.org/10.1002/cjce.24095

Oh, S. T. (2010). *Trend of mathematical models in microbial fuel cell for environmental energy refinery from waste/water* (pp. 25–30). https://doi.org/10.1007/978-3-642-13624-5_3

Ortiz-Martínez, V. M., Salar-García, M. J., de los Ríos, A. P., Hernández-Fernández, F. J., Egea, J. A., & Lozano, L. J. (2015). Developments in microbial fuel cell modeling. *Chemical Engineering Journal, 271*, 50–60https://doi.org/10.1016/j.cej.2015.02.076

Ortiz-Martínez, V. M., Salar-García, M. J., los Ríos, A. P., Hernández-Fernández, F. J., Sánchez-Segado, S., Egea-Larrosa, J. A., & Lozano-Blanco, L. J. (2014). Avances recientes en modelado y simulación de pilas de combustible microbianas. *DYNA-Ingeniería e Industria, 89*(6).

Ou, S., Kashima, H., Aaron, D. S., Regan, J. M., & Mench, M. M. (2017). Full cell simulation and the evaluation of the buffer system on air-cathode microbial fuel cell. *Journal of Power Sources, 347*, 159–169. https://doi.org/10.1016/j.jpowsour.2017.02.031

Owusu, P. A., & Asumadu-Sarkodie, S. (2016). A review of renewable energy sources, sustainability issues and climate change mitigation. *Cogent Engineering, 3*(1), 1167990. https://doi.org/10.1080/23311916.2016.1167990

Ozgun, H., Dereli, R. K., Ersahin, M. E., Kinaci, C., Spanjers, H., & van Lier, J. B. (2013). A review of anaerobic membrane bioreactors for municipal wastewater treatment: Integration options, limitations and expectations. *Separation and Purification Technology, 118*, 89–104. https://doi.org/10.1016/J.SEPPUR.2013.06.036

Palatsi, J., Ripoll, F., Benzal, A., Pijuan, M., & Romero-Güiza, M. S. (2021). Enhancement of biological nutrient removal process with advanced process control tools in full-scale wastewater treatment plant. *Water Research, 200*, 117212. https://doi.org/10.1016/j.watres.2021.117212

Pan, H., Chen, X.-Z., Liang, X.-F., Zhu, L.-T., & Luo, Z.-H. (2016). CFD simulations of gas–liquid–solid flow in fluidized bed reactors—A review. *Powder Technology, 299*, 235–258. https://doi.org/10.1016/j.powtec.2016.05.024

Panneerselvam, R., Savithri, S., & Surender, G. D. (2007). CFD based investigations on hydrodynamics and energy dissipation due to solid motion in liquid fluidised bed. *Chemical Engineering Journal, 132*(1–3), 159–171. https://doi.org/10.1016/j.cej.2007.01.042

Partopour, B., & Dixon, A. G. (2016). Reduced microkinetics model for computational fluid dynamics (CFD) simulation of the fixed-bed partial oxidation of ethylene. *Industrial & Engineering Chemistry Research, 55*(27), 7296–7306. https://doi.org/10.1021/acs.iecr.6b00526

Patziger, M. (2021). Improving wastewater treatment plant performance by applying CFD models for design and operation: Selected case studies. *Water Science and Technology, 84*(2), 323–332. https://doi.org/10.2166/wst.2021.019

Peleka, E. N., & Matis, K. A. (2016). Hydrodynamic aspects of flotation separation. *Open Chemistry, 14*(1), 132–139. https://doi.org/10.1515/chem-2016-0014

Pereira, J. P., Karpinska, A., Gomes, P. J., Martins, A. A., Dias, M. M., Lopes, J. C. B., & Santos, R. J. (2012). Activated sludge models coupled to CFD simulations. In *Single and two-phase flows on chemical and biomedical engineering* (pp. 153–173). Bentham Science Publishers Ltd. https://doi.org/10.2174/978160805295011201010153

Pilarek, M., Sobieszuk, P., Wierzchowski, K., & Dąbkowska, K. (2018). Impact of operating parameters on values of a volumetric mass transfer coefficient in a single-use bioreactor with wave-induced agitation. *Chemical Engineering Research and Design, 136*, 1–10. https://doi.org/10.1016/j.cherd.2018.04.012

Popat, S. C., & Torres, C. I. (2016). Critical transport rates that limit the performance of microbial electrochemistry technologies. *Bioresource Technology, 215*, 265–273. https://doi.org/10.1016/j.biortech.2016.04.136

Radeef, A. Y., & Ismail, Z. Z. (2019). Polarization model of microbial fuel cell for treatment of actual potato chips processing wastewater associated with power generation. *Journal of Electroanalytical Chemistry, 836*, 176–181. https://doi.org/10.1016/j.jelechem.2019.02.001

Ratkovich, N., Horn, W., Helmus, F. P., Rosenberger, S., Naessens, W., Nopens, I., & Bentzen, T. R. (2013). Activated sludge rheology: A critical review on data collection and modelling. *Water Research, 47*(2), 463–482. https://doi.org/10.1016/j.watres.2012.11.021

Ren, J., & Cao, S.-J. (2020). Development of self-adaptive low-dimension ventilation models using OpenFOAM: Towards the application of AI based on CFD data. *Building and Environment, 171*, 106671. https://doi.org/10.1016/j.buildenv.2020.106671

Ren, P., Li, W., & Yu, K. (2021). Computational fluid dynamics simulation of adsorption process in a liquid-solids fluidized bed. *Journal of Environmental Chemical Engineering, 9*(4), 105428. https://doi.org/10.1016/j.jece.2021.105428

Rivera-Alvarez, I., Brown, R. K., Keskin-Pyttel, D., Steffens, J., Farber, P., & Schröder, U. (2020). Correlating theoretical boundary layer thickness to the power output of a microbial fuel cell with a complex anode geometry operated at varying flow rates. *Journal of Power Sources, 470*, 228428. https://doi.org/10.1016/j.jpowsour.2020.228428

Runchal, A. K., & Rao, M. M. (2020). CFD of the Future: Year 2025 and Beyond. In *50 Years of CFD in Engineering Sciences* (pp. 779–795). Springer Singapore. https://doi.org/10.1007/978-981-15-2670-1_22

Saini, A. K., Paritosh, K., Singh, A. K., & Vivekanand, V. (2021). CFD approach for pumped-recirculation mixing strategy in wastewater treatment: Minimizing power consumption, enhancing resource recovery in commercial anaerobic digester. *Journal of Water Process Engineering, 40*, 101777. https://doi.org/10.1016/j.jwpe.2020.101777

Salehpour, R., Jalilnejad, E., Nalband, M., & Ghasemzadeh, K. (2020). Hydrodynamic behavior of an airlift reactor with net draft tube with different configurations: Numerical evaluation using CFD technique. *Particuology, 51*, 91–108. https://doi.org/10.1016/j.partic.2019.09.005

Samstag, R. W., Ducoste, J. J., Griborio, A., Nopens, I., Batstone, D. J., Wicks, J. D., Saunders, S., Wicklein, E. A., Kenny, G., & Laurent, J. (2016a). CFD for wastewater treatment: An overview. *Water Science and Technology, 74*(3), 549–563. https://doi.org/10.2166/wst.2016.249

Samstag, R. W., Ducoste, J. J., Griborio, A., Nopens, I., Batstone, D. J., Wicks, J. D., Saunders, S., Wicklein, E. A., Kenny, G., & Laurent, J. (2016b). CFD for wastewater treatment: An overview. In *Water Science and Technology* (Vol. 74, Issue 3, pp. 549–563). IWA Publishing. https://doi.org/10.2166/wst.2016.249

Sánchez, F., Rey, H., Viedma, A., Nicolás-Pérez, F., Kaiser, A. S., & Martínez, M. (2018). CFD simulation of fluid dynamic and biokinetic processes within activated sludge reactors under intermittent aeration regime. *Water Research, 139*, 47–57. https://doi.org/10.1016/j.watres.2018.03.067

Sandhibigraha, S., Sasmal, S., Bandyopadhyay, T. K., & Bhunia, B. (2019). Computational fluid dynamics analysis of flow through immobilized catalyzed packed bed reactor for removal of

4-chlorophenol from wastewater. *Environmental Engineering Research, 25*(6), 878–889. https://doi.org/10.4491/eer.2019.184
Sangeetha, T., Li, I.-T., Lan, T.-H., Wang, C.-T., & Yan, W.-M. (2021). A fluid dynamics perspective on the flow dependent performance of honey comb microbial fuel cells. *Energy, 214*, 118928. https://doi.org/10.1016/j.energy.2020.118928
Sathe, S. M., Chakraborty, I., & Ghangrekar, M. M. (2021). Wastewater treatment and concomitant bioelectricity production using microbial fuel cell: Present aspects, up-scaling and future inventiveness. *Transactions of the Indian National Academy of Engineering, 6*(3), 633–651. https://doi.org/10.1007/s41403-021-00245-8
Satpathy, K., Rehman, U., Cools, B., Verdickt, L., Peleman, G., & Nopens, I. (2020). CFD-based process optimization of a dissolved air flotation system for drinking water production. *Water Science and Technology, 81*(8), 1668–1681. https://doi.org/10.2166/wst.2020.028
Schwedhelm, I., Zdzieblo, D., Appelt-Menzel, A., Berger, C., Schmitz, T., Schuldt, B., Franke, A., Müller, F.-J., Pless, O., Schwarz, T., Wiedemann, P., Walles, H., & Hansmann, J. (2019). Automated real-time monitoring of human pluripotent stem cell aggregation in stirred tank reactors. *Scientific Reports, 9*(1), 12297. https://doi.org/10.1038/s41598-019-48814-w
Shadloo, M. S., Oger, G., & Le Touzé, D. (2016). Smoothed particle hydrodynamics method for fluid flows, towards industrial applications: Motivations, current state, and challenges. *Computers and Fluids, 136*, 11–34. https://doi.org/10.1016/j.compfluid.2016.05.029
Shen, L., Wu, Q., Ye, Q., Lin, H., Zhang, J., Chen, C., Yue, R., Teng, J., Hong, H., & Liao, B.-Q. (2023). Superior performance of a membrane bioreactor through innovative in-situ aeration and structural optimization using computational fluid dynamics. *Water Research*, 120353https://doi.org/10.1016/j.watres.2023.120353
Shen, R., Jiao, Z., Parker, T., Sun, Y., & Wang, Q. (2020). Recent application of Computational Fluid Dynamics (CFD) in process safety and loss prevention: A review. *Journal of Loss Prevention in the Process Industries, 67*, 104252. https://doi.org/10.1016/j.jlp.2020.104252
Shen, Z., Zhang, M., Fan, X., Shi, S., & Han, D. (2019). Hydrodynamic and flotation kinetic analysis of a large scale mechanical agitated flotation cell with the typical impeller and the arc impeller. *Minerals*, *9*(2). https://doi.org/10.3390/min9020079
Shi, Y., Wang, Z., Du, X., Gong, B., Jegatheesan, V., & Haq, I. U. (2021). Recent Advances in the Prediction of Fouling in Membrane Bioreactors. *Membranes, 11*(6), 381. https://doi.org/10.3390/membranes11060381
Shirazian, S., Rezakazemi, M., Marjani, A., & Moradi, S. (2012). Hydrodynamics and mass transfer simulation of wastewater treatment in membrane reactors. *Desalination, 286*, 290–295. https://doi.org/10.1016/j.desal.2011.11.039
Singh, H., & Hutmacher, D. W. (2009). *Bioreactor studies and computational fluid dynamics* (pp. 231–249). https://doi.org/10.1007/10_2008
Singhal, A., Cloete, S., Radl, S., Quinta-Ferreira, R., & Amini, S. (2017). Heat transfer to a gas from densely packed beds of monodisperse spherical particles. *Chemical Engineering Journal, 314*, 27–37. https://doi.org/10.1016/j.cej.2016.12.124
Sobieszuk, P., Zamojska-Jaroszewicz, A., & Makowski, Ł. (2017). Influence of the operational parameters on bioelectricity generation in continuous microbial fuel cell, experimental and computational fluid dynamics modelling. *Journal of Power Sources, 371*, 178–187. https://doi.org/10.1016/j.jpowsour.2017.10.032
Song, H.-S., Cannon, W., Beliaev, A., & Konopka, A. (2014). Mathematical modeling of microbial community dynamics: a methodological review. *Processes, 2*(4), 711–752. https://doi.org/10.3390/pr2040711
Song, X., Luo, W., Hai, F. I., Price, W. E., Guo, W., Ngo, H. H., & Nghiem, L. D. (2018). Resource recovery from wastewater by anaerobic membrane bioreactors: Opportunities and challenges. *Bioresource Technology, 270*, 669–677. https://doi.org/10.1016/j.biortech.2018.09.001
Suh, J.-W., Kim, J.-W., Choi, Y.-S., Kim, J.-H., Joo, W.-G., & Lee, K.-Y. (2018). Development of numerical Eulerian-Eulerian models for simulating multiphase pumps. *Journal of Petroleum Science and Engineering, 162*, 588–601. https://doi.org/10.1016/j.petrol.2017.10.073

Sultan, T., Ahmad, Z., Hayat, K., & Chaudhry, I. A. (2022). Computational analysis of three lamp close conduit water disinfection UV reactor. *International Journal of Environmental Science and Technology, 19*(5), 4393–4406. https://doi.org/10.1007/s13762-021-03344-9

Tan, J., Ji, Y.-N., Deng, W.-S., & Su, Y.-F. (2021). Process intensification in gas/liquid/solid reaction in trickle bed reactors: A review. *Petroleum Science, 18*(4), 1203–1218. https://doi.org/10.1016/j.petsci.2021.07.007

Trifi, D., Climent, J., Arnau, R., Carratalà, P., García, M., Beltrán, I., Badenes, C., Chiva, S., & Martínez-Cuenca, R. (2023). *Design and implementation of a passive agitator to increase UV dose in WWTPs disinfection channels* (pp. 626–636). https://doi.org/10.1007/978-3-031-15928-2_55

Wang, K., Li, L., Tan, F., & Wu, D. (2018). Treatment of landfill leachate using activated sludge technology: A review. *Archaea, 2018*, 1–10. https://doi.org/10.1155/2018/1039453

Wang, Y., Brannock, M., Cox, S., & Leslie, G. (2010). CFD simulations of membrane filtration zone in a submerged hollow fibre membrane bioreactor using a porous media approach. *Journal of Membrane Science, 363*(1–2), 57–66. https://doi.org/10.1016/j.memsci.2010.07.008

Wang, Y., Chen, J., & Larachi, F. (2013a). Modelling and simulation of trickle-bed reactors using computational fluid dynamics: A state-of-the-art review. *The Canadian Journal of Chemical Engineering, 91*(1), 136–180. https://doi.org/10.1002/cjce.20702

Wang, Y., Waite, T. D., & Leslie, G. L. (2013). Computational fluid dynamics (CFD) analysis of membrane reactors: Modelling of membrane bioreactors for municipal wastewater treatment. In *Handbook of Membrane Reactors* (Vol. 1, pp. 532–568). Elsevier Inc. https://doi.org/10.1533/9780857097330.3.532

Werner, S., Kaiser, S. C., Kraume, M., & Eibl, D. (2014). Computational fluid dynamics as a modern tool for engineering characterization of bioreactors. *Pharmaceutical Bioprocessing, 2*(1), 85–99.

Weyand, B., Israelowitz, M., Kramer, J., Bodmer, C., Noehre, M., Strauss, S., Schmälzlin, E., Gille, C., von Schroeder, H. P., Reimers, K., & Vogt, P. M. (2015). Three-dimensional modelling inside a differential pressure laminar flow bioreactor filled with porous media. *BioMed Research International, 2015*, 1–9. https://doi.org/10.1155/2015/320280

Wicklein, E., Batstone, D. J., Ducoste, J., Laurent, J., Griborio, A., Wicks, J., Saunders, S., Samstag, R., Potier, O., & Nopens, I. (2016). Good modelling practice in applying computational fluid dynamics for WWTP modelling. *Water Science and Technology, 73*(5), 969–982. https://doi.org/10.2166/wst.2015.565

Wu, B. (2013). Advances in the use of CFD to characterize, design and optimize bioenergy systems. *Computers and Electronics in Agriculture, 93*, 195–208. https://doi.org/10.1016/j.compag.2012.05.008

Wu, B., & Chen, S. (2008). CFD simulation of non-Newtonian fluid flow in anaerobic digesters. *Biotechnology and Bioengineering, 99*(3), 700–711. https://doi.org/10.1002/bit.21613

Wu, Q., Yan, X., Xiao, K., Guan, J., Li, T., Liang, P., & Huang, X. (2018). Optimization of membrane unit location in a full-scale membrane bioreactor using computational fluid dynamics. *Bioresource Technology, 249*, 402–409. https://doi.org/10.1016/j.biortech.2017.09.209

Xanthos, S., Gong, M., Ramalingam, K., Fillos, J., Deur, A., Beckmann, K., & McCorquodale, J. A. (2011). Performance assessment of secondary settling tanks using CFD modeling. *Water Resources Management, 25*(4), 1169–1182. https://doi.org/10.1007/s11269-010-9620-1

Xia, C., Zhang, D., Pedrycz, W., Zhu, Y., & Guo, Y. (2018). Models for microbial fuel cells: A critical review. *Journal of Power Sources, 373*, 119–131. https://doi.org/10.1016/j.jpowsour.2017.11.001

Xiao, K., Liang, S., Wang, X., Chen, C., & Huang, X. (2019). Current state and challenges of full-scale membrane bioreactor applications: A critical review. *Bioresource Technology, 271*, 473–481. https://doi.org/10.1016/j.biortech.2018.09.061

Xiao, Y., Peng, J., Sun, W., Han, H., Xie, L., & Wang, R. (2023). Study on liquid-solid fluidisation behaviour of coarse coal particles: Experiment and CFD simulation. *Chemical Engineering Research and Design, 189*, 1–13. https://doi.org/10.1016/j.cherd.2022.11.001

Xie, F., Ge, H., Liu, J., Chen, W., & Song, H. (2018). CFD and experimental studies the effect of micro-channel turbulence promoter installation on the hydrodynamic performance of submerged flat-sheet membrane bioreactor. *Chemical Engineering and Processing - Process Intensification, 127*, 28–35. https://doi.org/10.1016/j.cep.2018.03.009

Xu, C., Rangaiah, G. P., & Zhao, X. S. (2015). A computational study of the effect of lamp arrangements on the performance of ultraviolet water disinfection reactors. *Chemical Engineering Science, 122*, 299–306. https://doi.org/10.1016/j.ces.2014.09.041

Xu, H., Yang, B., Liu, Y., Li, F., Shen, C., Ma, C., Tian, Q., Song, X., & Sand, W. (2018). Recent advances in anaerobic biological processes for textile printing and dyeing wastewater treatment: A mini-review. *World Journal of Microbiology and Biotechnology, 34*(11), 165. https://doi.org/10.1007/s11274-018-2548-y

Xu, N., Fan, L., Pang, H., & Shi, H. (2010). Feasibility study and CFD-aided design for a new type oxidation ditch based on airlift circulation. *The Canadian Journal of Chemical Engineering, n/a-n/a*. https://doi.org/10.1002/cjce.20335

Xu, Q., Wan, Y., Wu, Q., Xiao, K., Yu, W., Liang, S., Zhu, Y., Hou, H., Liu, B., Hu, J., Yang, Y., & Yang, J. (2021). An efficient hydrodynamic-biokinetic model for the optimization of operational strategy applied in a full-scale oxidation ditch by CFD integrated with ASM2. *Water Research, 193*, 116888. https://doi.org/10.1016/j.watres.2021.116888

Yang, J., Wu, J., Zhou, L., & Wang, Q. (2016). Computational study of fluid flow and heat transfer in composite packed beds of spheres with low tube to particle diameter ratio. *Nuclear Engineering and Design, 300*, 85–96. https://doi.org/10.1016/j.nucengdes.2015.10.030

Yao, Y., Fringer, O. B., & Criddle, C. S. (2022). CFD-accelerated bioreactor optimization: Reducing the hydrodynamic parameter space. *Environmental Science: Water Research & Technology, 8*(3), 456–464. https://doi.org/10.1039/D1EW00666E

Ye, C.-C., Zhang, P.-J.-Y., Wan, Z.-H., Yan, R., & Sun, D.-J. (2022). Accelerating CFD simulation with high order finite difference method on curvilinear coordinates for modern GPU clusters. *Advances in Aerodynamics, 4*(1), 7. https://doi.org/10.1186/s42774-021-00098-3

Yi, Y., Xie, B., Zhao, T., Qian, Z., & Liu, H. (2020). The effect of anode hydrodynamics on the sensitivity of microbial fuel cell based biosensors and the biological mechanism. *Bioelectrochemistry, 132*, 107351. https://doi.org/10.1016/j.bioelechem.2019.107351

Zawawi, M. H., Saleha, A., Salwa, A., Hassan, N. H., Zahari, N. M., Ramli, M. Z., & Muda, Z. C. (2018). *A review: Fundamentals of computational fluid dynamics (CFD)*. 020252. https://doi.org/10.1063/1.5066893

Zeng, Y., Choo, Y. F., Kim, B.-H., & Wu, P. (2010). Modelling and simulation of two-chamber microbial fuel cell. *Journal of Power Sources, 195*(1), 79–89. https://doi.org/10.1016/j.jpowsour.2009.06.101

Zhan, M., You, M., Liu, L., Zhang, Y., Yuan, F., Guo, B., Cheng, G., & Xu, W. (2021). Numerical simulation of mechanical flocculation in water treatment. *Journal of Environmental Chemical Engineering, 9*(4), 105536. https://doi.org/10.1016/J.JECE.2021.105536

Zhang, M., Dong, H., & Geng, Z. (2018). Computational study of flow and heat transfer in fixed beds with cylindrical particles for low tube to particle diameter ratios. *Chemical Engineering Research and Design, 132*, 149–161. https://doi.org/10.1016/j.cherd.2018.01.006

Zhang, S., Müller, D., Arellano-Garcia, H., & Wozny, G. (2013). CFD Simulation of the Fluid Hydrodynamics in a Continuous Stirred-Tank Reactor. In *Chemical Engineering Transactions* (Vol. 32). https://doi.org/10.3303/CET1332241

Zhao, L., Li, J., Battaglia, F., & He, Z. (2016). Investigation of multiphysics in tubular microbial fuel cells by coupled computational fluid dynamics with multi-order Butler-Volmer reactions. *Chemical Engineering Journal, 296*, 377–385. https://doi.org/10.1016/j.cej.2016.03.110

Chapter 4
Computational Fluid Dynamics in Biomedical Engineering

1 Introduction

The most recent developments in computational fluid dynamics (CFD) applications for studying its importance in the biomedical field are presented in this chapter. CFD can help you understand how changes in devices and formulations affect critical in vitro parameters. In order for the human body to operate properly, a complicated interaction between fluids and solid structures must take place. Fluids are crucial for the healthy functioning of the main systems of the human body such as respiratory, integumentary, cardiovascular, lymphatic, gastrointestinal, neurological, reproductive, or urine systems. Despite the complexity of human anatomy and the fluid nature of the human body, recent advances in computer science have generated outstanding performance hardware and software obtainable for use in biomedical fields, rendering them more affordable and practical. Research and findings from the laboratory on current initiatives are among the contributions. Hence more intricate models can be simulated using CFD simulations and findings from this work may be utilized to create an efficient experimental model. This chapter highlights a few of the important biomedical applications to enhance either the system design or to increase the efficiency of the outcome. To comprehend the physiology and pathophysiology of the human system through simulation, this chapter addresses the fundamental approach of CFD as a trustworthy tool for researchers and medical scientists. Before making a serious commitment to carry out any medical design adjustments, CFD plays a significant function in decision support and gives guidance for developing medical treatments. Future uses of CFD modelling include reducing the need for human subject testing when creating new devices and formulations and establishing bioequivalence to expedite the approval of new devices. It offers new knowledge about the dynamics of disease or devices that will eventually lead to the development of a map of novel treatment strategies.

S. E. Jujjavarapu et al., *Computational Fluid Dynamics Applications in Bio and Biomedical Processes*, https://doi.org/10.1007/978-981-99-7129-9_4

2 Background

Biomedical engineering is the activity of applying engineering concepts and methodologies for problem-solving in biology and medicine. Many of the challenges that health professionals face have historically captivated the attention of engineers because they involve procedures that are fundamental to engineering practice. Biomedical engineers use standard engineering approaches to understand, change, or govern biological activities, as well as to design and manufacture devices that may assist in the management and identification of human illnesses. Over the last decade, the convergence of medical imaging and computer modelling technology has enabled extraordinary advances in the development and use of computational fluid dynamics modelling of patient-specific problems.

CFD is a subfield of fluid dynamics that simulates real fluid flows by using numerical solutions to governing equations. It is one of the most extensively used computer-based simulation approaches for simulating actual fluid flows by using numerical solutions to governing equations. CFD is a widely used technique for computer-based simulation to address complicated issues in a variety of contemporary engineering domains, including biomedicine (Reid, 2021). To generate updated designs and optimize them through computer simulations, CFD is increasingly important. This leads to lower operating costs and more efficiency. Due to the relevance, utility, and fast implementation of computer-based simulation across a wide range of applications, significant advancements in this field are rising quickly. The use of CFD software in biomedical research has allowed simulations of the physiology and pathophysiology of biological systems.

CFD is being used more and more in a variety of essential engineering systems, involving both a branch of fluid mechanics and a specialized field of mathematics. CFD modelling has already received a lot of attention in the fields of biomedical research and the development of medical equipment. Additionally, a thorough characterization of complex physiology and the evaluation of computation metrics may be established by combining imaging approaches with CFD simulation. Doctors may employ CFD models as therapeutic tools to address a variety of various cardiovascular and respiratory problems. Therefore, this research explores the CFD study using the most recent advancements in the clinical field with an emphasis on biomedical applications (Bluestein, 2017; Tsega, 2018). In computational results, the relationship between engineering and medical concepts is advantageous.

The design of medical equipment and the study of physiological events are only two examples of the many research issues in medicine and biology that have been made clear by the use of CFD modelling (Sotiropoulos, 2012; Zhong et al., 2018). The advancements in biomedical computing have changed how biology and medicine are studied and administered. Biomedical computing offers a wide range of benefits. They are essential for generating novel clinical therapies, advancing exciting new biomedical discoveries, and improving our understanding of human physiology. Numerous numerical techniques are used for simulation in the biomedical field. Methods like the finite element method (FEM) are utilized for structural analysis. Examples include analysing the stress distribution in hip implants or doing a stress analysis of the human skull during a head collision. Understanding fluid motion inside and around the body is made easier by computational fluid dynamics techniques. Understanding fluid dynamics within and around the body is aided by CFD techniques (Zamir et al., 2010). CFD is used in the biomedical area to study the flow of blood in arteries, simulate airflow within the respiratory tract, and so on. Thermal analysis is another intriguing tool for determining how heat is transmitted between various bodily parts and the environment (Yang et al., 2017). One such use is the thermal study of cooling a human heart during cardiac surgery. Multibody dynamics is a crucial simulation technique that aids in comprehending human biomechanics. Understanding the actions of a physically challenged or injured person is made easier by doing this. The most modern biomedical computing technology is the optimization technique. Medical gadgets that are implanted in patients' bodies are now designed better thanks to optimization algorithms. Examples of typical examples include the optimization of patient-specific hip implants and stents (Jamari et al., 2022; Prabowo et al., 2020). The modelling of patient-specific computational fluid CFD is a relatively new discovery.

Engineers have been analysing heat transfer and fluid flow phenomena using CFD for more than 50 years. Applications of CFD in biomedical and health research have recently advanced quickly. It has been applied to the development of medical devices, the evaluation of drug delivery systems, the analysis of physiological flows, and the facilitation of surgical planning. Because of the complicated nature associated with such fluid flows, a multidisciplinary strategy is needed for the development of computational tools and software to solve equations involving mathematics. CFD has grown more accessible and useful in a wider range of applications as a result of technical advancements and reducing computational costs. CFD is presently most commonly used in biomedical research in the respiratory, cardiovascular, muscular, and neural systems. CFD is also used in cerebral fluid, synovial joint, and intracellular fluid studies. Even though CFD can produce valuable and appealing findings, persons who do not have an extensive knowledge of mathematics and engineering principles may struggle to interpret the data.

As a result, the continued growth and development of computational medicine will necessitate considerable collaboration among experts in computer science, engineering, and biological research. This chapter attempts to solve knowledge gaps on the CFD technique method, and its applications in biomedical research. The procedures for CFD's modelling are categorized into three primary categories: preprocessor, solver (processor), and post-processor. The modelling element's input, known as the preprocessor, involves problem-solving, discretization (meshing), and the creation of a computational model. The processing component, known as the solver, uses algebraic and numerical methods to solve governing equations. The post-processor is the output component where the computational outcomes are represented visually by attaining a satisfactory convergence of the solved equations of state for each cell.

Pertaining to the below Figure (Fig. 1), the modelling element's input (the preprocessor) involves problem-solving, discretization (meshing), and the creation of a computational model. The region of interest's geometry must be specified, the grid or mesh must be generated, the physical, as well as chemical processes that have to be modelled, must be chosen, the fluid characteristics must be defined, and the suitable boundary conditions for the intake and outflow must be specified. The precision of the answer increases with the size of the cell grids. Grid fineness affects the precision of response and the time needed to solve a computer problem. Hence, this process takes up the majority of the time. As biological fluids have different viscosities depending on their shear rates, they behave differently than Newtonian fluids (Baieth, 2008). As a result, the range of shear rates should be taken into account while choosing the appropriate viscosity model using a mathematical calculation. Boundary conditions such as the pressure, flow rate, and temperature are conveniently accessible according to the region of interest. The reality is that such boundary conditions vary according to the functioning of the system under consideration (Lee, 2011).

The solver is a computational component, and it generates various forms of numerical results by using equations that govern and algebraic results. There are numerous numerical solution methods available, such as finite volume, finite element, and finite

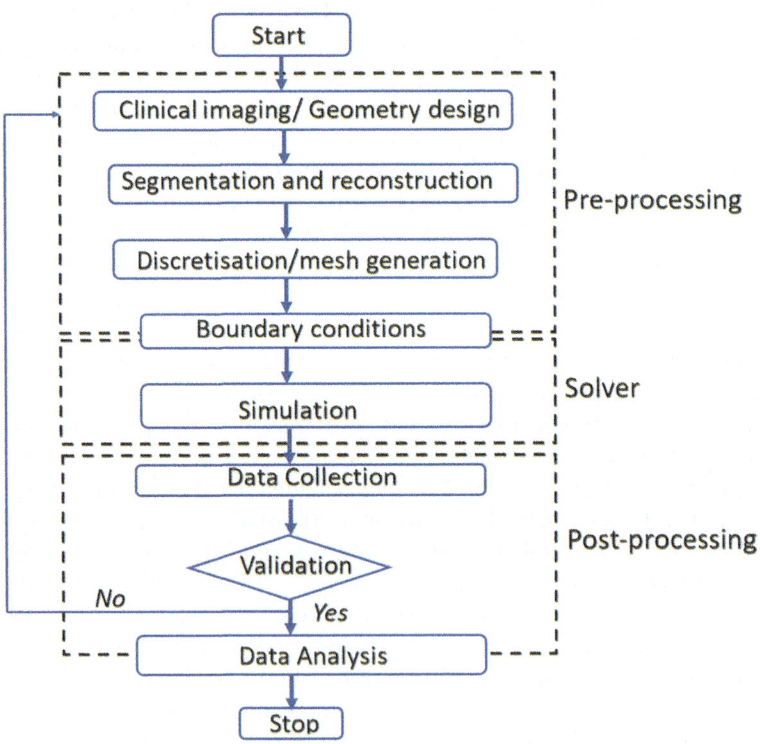

Fig. 1 Flowchart representing CFD simulation for biomedical applications

difference methods. Despite the fact that each solver takes a different numerical technique, they all begin by estimating the unidentified flow variables with simple functions, then discretizing the outcomes by replacing the findings into the governing flow and calculating the algebraic equation.

By attaining a satisfactory convergence from the solver, the post-processor is the component where the results of the simulations are visualized. The goal of this procedure is to visualize the outcomes of the computation. In addition to domain geometry and grid displays, visualization techniques such as vector plots, line and coloured contour plots, two-dimensional and three-dimensional surface plots, and particle tracking have been developed. Using this method, the researcher may easily interpret the simulation data. Variations in flow profiles, pressure distribution, wall shear stress (WSS), oscillation shear index (OSI), and shear rate, for example, can be observed using colour rendering techniques.

The researchers carried out extensive research on the utilization of ANSYS Fluent software to simulate a broad variety of issues in numerous CFD domains. Some of the research carried out in this field includes the simulation of blood flow in clogged arteries, which shows how blood behaves when there is a curved blockage in the centre of the blood flow stream as an indication of clogging (Pandey et al., 2020).

Theoretical gains are provided by CFD by focusing on the design and solution of the governing mathematical equations, as well as the examination of numerous approximations to these equations. Meanwhile, experimental and numerical approaches highlighted the importance of CFD as an affordable option for modelling real flow of fluids, especially within human body systems. As a result, when comparing the fluid dynamics of analytical and experimental methodologies, it provides detailed visual and comprehensive information. CFD is significant since it provides opportunities for simulation to make decisions to provide the right medical treatments and carry out any design changes. Due to the significance of computational medical models of circulatory systems, research into biomedical CFD applications has attracted a lot of interest recently. CFD is crucial since it provides opportunities for simulation before making a genuine commitment to create the right medical treatments and carry out any design changes. Due to the significance of computational medical models of circulatory systems, research into biomedical CFD applications has attracted a lot of interest recently. The next subsection goes over the biomedical CFD applications for biological systems.

This chapter comprises research on biomedical engineering topics, with a focus on the fundamental knowledge of the flow of fluids in biological systems. The four components of the whole modelling system are (a) development of geometric models; (b) generation of dynamic boundary conditions; (c) numerical simulation; and (d) visualization of flows. The next section describes the specifics of the procedures used for each component.

3 Reconstruction of the System

The first step in reconstructing detailed 3D views of the organs is image acquisition, which involves acquiring diagnostic pictures from computed tomography, also known as CT, or MRI imaging. As a result, a 3D matrix will be formed by combining many 2D cross-sections. This matrix contains details regarding tissues and anatomical structures. The area of interest needs to be segmented in order to recreate the respiratory tract using medical imaging. A uniform area is extracted during the segmentation method and is ready for three-dimensional processing. The three processes that can be utilized to replicate the human organ are capturing image, the process of segmentation and surface/volume reconstruction.

3.1 Acquisition of Medical Image

Image acquisition is the first stage in surface reconstruction. The process of gathering diagnostic pictures from diverse sources, like CT or MRI, is referred to as this stage.

3.1.1 Computed Tomography (CT)

Sir Godfrey Hounsfield invented computed tomography, also known as CT, in 1972, using a huge quantity of X-rays to create a sequence of planar cross-sectional views along an axis (Richmond, 2004). Hundreds of X-rays are blasted through the body amid a minimally invasive CT scan to create three-dimensional images that can be viewed (Inthavong et al., 2009).

Numerous X-rays are taken at extremely small cross-sections in the region of greatest interest over the subject's body to produce slices in a CT scan. During a scan, photon-containment X-rays get scattered by biological tissues and some lose the energy that they contain as a result. Electronic detectors capture data collected from every cross-X-ray segment, which is then sent to a computer system to be combined into one image. CT scan images are analogous to or superior to MRI images in terms of resolution. While CT is preferable for bone scanning and bone cancers, it does not provide soft-tissue contrast that MRI does.

The exposure to radiation may be an issue due to the type of this operation. The most basic CT scanner is made up of precise X-ray beam emitters and a precise radiation sensor. The patient rests flat on a table for a traditional CT scan. The patient is then moved by a revolving, toroidal gantry that contains a tube for X-rays and electronic X-ray detectors. Several X-rays in thin cross-sections across the person's torso are obtained, resulting in slices. The detectors collect information gathered from every cross-X-ray sector and transmit it to a computer, where it is combined into a single image. As the X-ray beam passes through the body, different anatomical characteristics scatter or absorb it, reducing the received X-ray beam.

3.1.2 Magnetic Resonance Imaging (MRI)

MRI scans provide cross-sectional pictures of a person's body by using energy from radiofrequency waves and magnetic fields (Sharp et al., 2010). A magnetic field is created during a scan by an electric current flowing via wire coils. The magnetic field is then aligned with the protons of hydrogen in the body's inner water molecules, which spin randomly. A brief blast of tunable radio waves strikes the body, briefly altering the quantum condition of the hydrogen protons. Once the radiofrequency wave ceases, the proton resumes to its initial location, emitted a radio signal which a scanner detects and analyses as images. It implies that various structures in tissues produce distinct pulse patterns, resulting in contrast disparities across tissue qualities. The major distinction is that nuclear magnetic resonance energy does not subject the person receiving treatment to ionizing radiation. MR imaging creates a three-dimensional image using fields of magnets and radio waves pulses rather than radiation, protecting the individual receiving treatment from ionizing radiation. A huge superconductive magnet, approximately 1.5 T, is housed inside the cylinder and is nearly fifty thousand times more powerful than Earth's own magnetic field. When activated, an electric current flow through the coils, producing both a powerful static magnetic field and a weak oscillating magnetic field.

The fundamental advantage of MRI is the lack of ionizing radiation. MRI offers an excellent contrast resolution as well, but it is a more expensive and slower approach. Because MRI does not count photons, no prior values may be used to calculate thresholds. This makes automating the rebuilding procedure more challenging. By substantially shifting the magnetic field, ferromagnetic materials impair MR. MRI scans are also more accurate than CT images in identifying structures. A 3D matrix composed of a series of two-dimensional cross-sections isolated by a distance contains information on tissues and structures distinguishable by variations in brightness or greyscale.

When a 3D surface is needed for simulations using CFD, even little changes to the three-dimensional surface can have a significant impact upon the resulting flow. As a result, image preprocessing will almost certainly be required in this case. There are several formats for the produced images from scanners, but DICOM is among the most common and popular because it includes patient data and the data associated with the images. When these photos, which are made up of layered two-dimensional pixels split as a slice thickness, are combined, they produce 3D volumetric data. Using grayscale intensities, the pixels in the photos show various tissues or organs. These data in an MRI or CT scan represent a mapping of radio density or the measurement of the linear X-ray attenuation coefficient. In contrast with image processing, in which the greyscale usually ranges from 0 to 255, medical imaging employs Hounsfield scaling using HU units.

A complete MRI/CT scan generates a series of two-dimensional cross-sectional pictures, typically separated by thickness. These photos can be combined to provide 3D volumetric data. Each image is composed of an image's pixel arrays, which refers to the smallest component of a two-dimensional image. Whenever the variation in thickness of slices among two images is taken into account, these pixels transform into voxels. As the accuracy of the spatial resolution obtained from scan is related to the voxel volume, higher resolution makes distinguishing among various anatomical components easier. Thus, matrices, which are just 2D or 3D arrays, can represent the digital version of a picture.

The next sections go through a few common segmentation algorithms utilizing threshold-based, edge detection-based, as well as region-based segmentation techniques in greater depth.

4 Image Segmentation

Image segmentation is a challenging and crucial phase of digital image processing. The precision and impact of segmentation have an impact on whether or not the computation and analytic process is ultimately successful (Steinman, 2002). As a result, attention should be made on basic segmentation of pictures algorithm research and progress within an extensive variety of applications. The area of interest has to be subdivided in order to replicate the morphological patterns from the diagnostic imaging series. These photos are divided into multiple homogeneous segments during segmentation, so that any two contiguous segments may end up in a heterogeneous segment (Bertolini et al., 2022). A voxel is the natural three-dimensional expansion of an image pixel and the core unit of the widely used DICOM medical imaging formatting standard. In reality, slices—a group of aligned scanned images—can be utilized to analyse voxel-based data (Temor et al., 2022). A voxel is a volume element formed by each cross-sectional slice in a two-dimensional slice separated by a defined thickness from the next. A zone inside an image can be identified by its pixel attributes (for example, grayscale intensity) or boundary. As a result, segmentation algorithms can be loosely classified into three categories (Bali et al., 2015): edge-based detection, thresholding-based detection, region-based detection, and deep learning-based segmentation which are discussed further in the next section.

4.1 Edge Detection

Edge-based segmentation builds a boundary across the area of interest by recognizing edge pixels. This approach works best on images of anatomical elements with well-defined borders, such as arteries. However, noise can commonly result in incorrect or missed recognition of edges, which is a common problem. If the boundaries of a certain structure can be recognized, the contained area containing that structure can be retrieved and segmented from the scan. This border detection approach is an edge-based segmentation method. Edge-based segmentation refers to a wide range of techniques that are based on the idea that an edge can be defined as a region in the image wherein the intensity changes fast. This is accomplished by applying a filter to an image that recognizes the abrupt variation in the value of pixels and classifies the pixel as relating to or not relating to an edge. Edge detection maintains an image's critical structural traits by removing extraneous information and significantly lowering the amount of data to be processed (Dhankhar et al. 2013). There are several edge detection techniques; however, they can be categorized into two types:

- Derivative-based, in which the size of the very first derivative of the intensity change goes over a predetermined threshold;
- Gradient-based, where a second derivative of the intensity exceeds a zero line (i.e., a point where the sign of a function changes from positive to negative, symbolized by the crossing of an axis with a zero value). During gradient-based edge detection, a gradient of successive pixels is generated in both the x- and y-directions.

4.2 Thresholding-Based Segmentation

It has evolved into the most fundamental and extensively used segmentation method in picture segmentation because of its straightforward implementation, inexpensive computation, and steady performance (Heet al., 2022). Threshold segmentation is often sometimes referred to as the "image binarization procedure" or pixel-based segmentation since the resultant picture typically only contains two grey values: 255 and 0. The foreground and background may be distinguished using the threshold segmentation technique. Threshold segmentation is particularly helpful for segmenting photos with a large contrast between the foreground items and background since it primarily recovers the foreground based on grey value information. Enhancing the overall contrast of the pictures is required before doing threshold processing to segment images with quite low contrast (Niu et al. 2019).

Two common threshold segmentation techniques are:

(a) Global threshold segmentation: Global threshold segmentation classifies pixels as white or black based on whether or not the grey value of the pixel is more or less than the threshold value. Alternatively, pixels higher than that of threshold are converted to black, whereas those that are smaller or equivalent to the threshold are assigned to white.
(b) Adaptive local threshold segmentation: Adaptive thresholding generates different threshold values for different locations by estimating the threshold amount for smaller regions.

4.3 Region-Based Segmentation Technique

In contrast to threshold-based segmentation, which employed values for thresholds driven by the intensity of pixel values to identify regions, edge-based segmentation involves determining the edge borders among sections based on pixel differences (Dar et al. 2019). This section will explain region-based segmentation which directly decides the region of interest. By integrating similar nearby pixels, region-based segmentation takes into account the pixel grayscale levels from nearby pixels. Region-based segmentation methods are among the most comprehensive yet challenging techniques since regions of interest provide more pixel classification than edges. Region-expanding algorithms are also helpful in noisy images with hard-to-see edges. Region-based segmentation determines the regions directly from within and expanding outward, as opposed to analysing from the outside in (Gould et al., 2009). The challenges are caused by the region classification criteria, which are frequently more difficult than edge detection methods. Additionally, it frequently fails to find objects that span several disconnected zones.

4.4 Deep Learning-Based Segmentation

Due to the success of deep learning models in a range of vision applications, there have been various studies aimed at developing image segmentation algorithms using these models (Guo et al., 2019). In this segmentation procedure, a neural network is given a series of images that have already been segmented using a deep learning methodology, and the algorithm creates the best way to predict further segmentation. It takes a lot of segmented picture datasets to train a neural network (Minaee et al., 2021). Once the output file has been segmented, it may be read directly into a CFD meshing program.

5 Mesh Generation

The next phase is mesh creation, also known as grid generation or discretization, which entails breaking the domain up into smaller sections known as subdomains (Dyedov et al., 2009). Prior to computational field simulation, the domain of interest must be discretized. Mesh creation plays a crucial role in the modelling process since it affects both the simulation's duration and the analysis's precision (Lintermann, 2021). Small cells are made to fill the volume; each cell is going to represent a different region that depicts the local flow. Then, each mesh cell is subjected to mathematical equations relating to flow physics. The mesh elements or cells are typically hexahedral, rectangular, pyramids, triangles, wedges, or triangular prisms, or polyhedral since the nasal airway volume is considered as a 3D geometry. It is helpful to think of the topology of a mesh as a hierarchy, with higher topology assuming that lower topologies exist. Building a top-notch mesh is crucial to obtaining accurate results and encouraging numerical stability. There are several methods for creating grids, and numerous algorithms have seen advancements that have made it possible to produce high-quality grids quickly. All techniques can be roughly categorized into three categories: Cartesian meshing, multiblock meshing, and unstructured meshing.

6 Boundary Condition Generation

After generating the standard grid, the following step is to define the border and domain. The boundary tags require labelling the domain's surfaces with labels such as inlet, outlet, or wall. The domain tags are used to classify the volume as solid or fluid. The form, size, and value of a known flow characteristic on the border determine the individuality of boundary conditions. Von Neumann and/or Dirichlet conditions are usually recommended for inlets and outlets. A von Neumann condition is used to extrapolate the density from the subsequent inner computational elements and a Dirichlet condition to specify the velocity. Inlets frequently use this to build sources. A straightforward outlet condition uses a von Neumann condition for the velocity and a Dirichlet condition for the density (Lintermann, 2021).

7 Applications of CFD in Biomedical Applications

In the biomedical profession, the use of CFD is frequently employed to resolve challenging issues. In order to generate updated designs and optimizations using computer simulations, CFD is increasingly important. This leads to lower operating costs and higher efficiency. To examine the analyses in biomedical applications, notably in blood flow and nasal airflow, several simulations and clinical results have been used. The analysis of blood flow covers the research of heart valves, coronary arteries, and blood circulation in ventricles. Also, the nasal airflow analysis includes improvements to drug administration, virtual surgery, and basic airflow in a nose. The next section discusses the use of CFD in biomedical applications:

7.1 Cerebral Aneurysm

The deformation of a section within the cerebral artery is a common cerebrovascular disease (Ishida et al., 2021). The study of pathological aneurysmal properties, including rupture, growth, and clinical effects, is increasingly frequently conducted using CFD simulation (Fig. 2). Utilizing the patient-specific geometry model, Steinman and co-authors used CFD for a brain aneurysm in the beginning (Lim et al., 2020; Steinman et al., 2003). CFD simulations are specifically used to look into the connections between hemodynamic factors and aneurysm features and consequences. The use of patient-specific parameters is necessary for a more accurate representation of the aneurysm hemodynamics. Clinical pictures of the patient, such as magnetic resonance scans or angiograms, can be used to determine the flow velocity parameters to be utilized for simulations. One can recreate the three-dimensional geometry particular to the patient and simulate the hemodynamic environment. This will make it possible to replicate the hemodynamic conditions linked to aneurysmal disease and the accompanying clinical outcomes will be more accurate and realistic. To characterize the aneurysm: the flow field and hemodynamic parameters like velocity, pressure, vortices, and wall shear stress (WSS) can be taken into consideration (Shishir et al., 2015).

CFD procedure for cerebral aneurysms: The Mimics Innovation Suite program is capable of extracting aneurysm morphology in stereolithography, or (.stl) format from 3D rotational and 3D CT angiography DICOM images. 3-Matic software can segment geometry in the area of interest from the .stl file. Mathematical modelling can be done using ANSYS software. For the fluid domain offered for 3D laminar flow fields, the Navier–Stokes and continuity formulas will be used, which can subsequently be discretized utilizing the finite volume method. By adapting the typical flow waveform derived from phase-contrast MRI imaging to the inlet, it is able to calculate the flow rate required to provide a physiological wall shear stress (WSS).

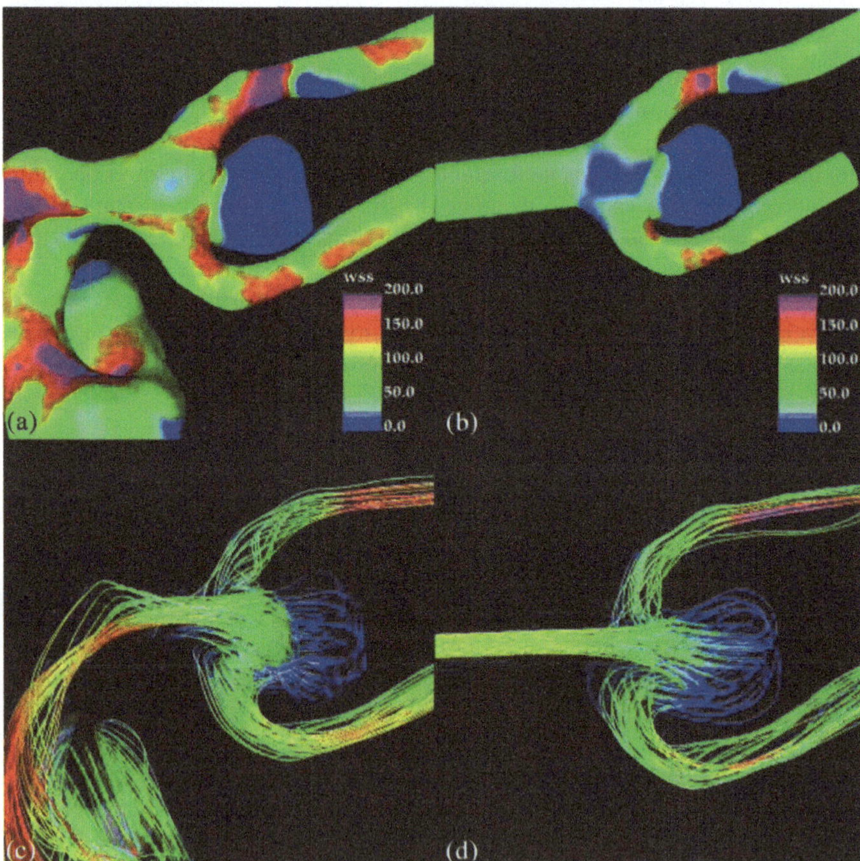

Fig. 2 Wall shear stress observed during simulation of intra-aneurysmal hemodynamics (Castro et al., 2006)

Uchiyama and co-authors utilized CFD to study the hemodynamic environment of an aneurysm by mimicking the vascular configuration of a patient who had two overlapping flow diverters (FDs) by using two stent patterns. Various parameters such as mass flow rate, wall shear stress, and velocity were calculated for each pattern and compared to the results of the no-FD pattern. According to the results of the CFD simulation, the deployment of such a single FD had a minimal effect on the properties of the flow of blood inside the aneurysm; however, the outcomes of the overlapping FD patterns revealed significant changes in the flow. The flow structure observed under the single-FD pattern conditions was similar to the flow structure observed under no-FD conditions. As per their results, the overlapping pattern demonstrated much lower flow velocity and WSS as compared to the non-stent sample (Uchiyama et al., 2021). With the advancement in research, the aim is to put the focus on the necessity of thorough multimodality hemodynamic metric assessment. The researchers utilized patient-specific aneurysm models and performed pulsatile volumetric particle velocimetry experiments. With the in vivo 4D flow MRI acting as boundary conditions for the CFD and particle velocimetry, they used a novel multimodality technique that combines in vivo measurements, in vitro experimentation, and in silico modelling (Brindise et al., 2019).

7.2 Respiratory Tract

From the nasal cavity to the alveoli, the human respiration system resembles an inverted tree (Johnston et al. 2008). The basic function of the respiratory system is to transport gases such as carbon dioxide and oxygen between the atmosphere to the lungs (Narendrakumar et al. 2022). The flow fields of the small bronchial tubes are generally laminar. But because of the geometrical consequences of the lung airways, they are rather complicated. The primary characteristics of the lung airway architecture include asymmetry, nonplanarity, and many generations. In the human lung, there are 23 generations of airways (Weibel et al. 1962). With each successive generation, the intricacy of the airflow in the lung airways grows. Modelling flow and deposition of particles in multigenerational bronchial tube geometries is highly difficult due to the geometry's complexities and the magnitude of the problem. The unpredictability of inhalation-exhalation respiratory cycles complicates matters. Flow and deposition of particles in an ideal lung shape are simulated for the investigation. Using the CFD method, the researchers modelled airflow and aerosol-particle accumulation in the individual's respiratory system (Fig. 3). Based on the Weibel model, they included third through sixth-generation bronchus in the three-dimensional respiratory systems and carried out computation under the assumption of laminar flow. Discrete-phase modelling is used to investigate the two-phase flow. The outcomes of velocity contours, aerosol-particle deposition efficiency, and aerosol-particle entrapment processes were evaluated at various places across the respiratory tract (Srivastav et al., 2011).

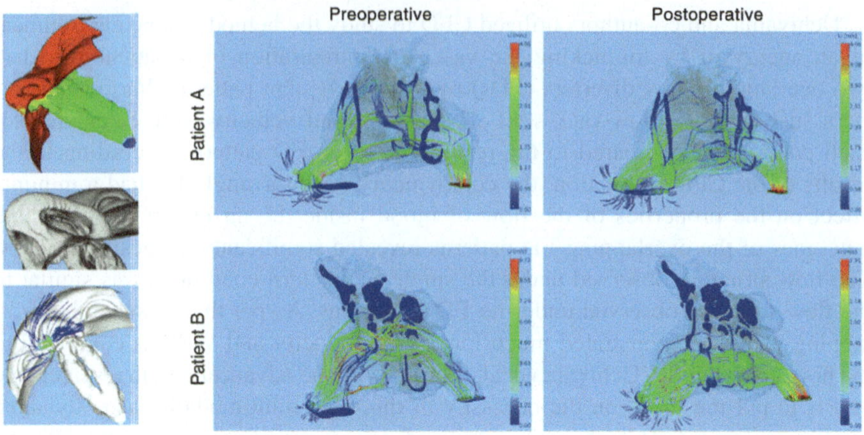

Fig. 3 Pressure profile studied during nasal airflow via CFD simulation and visualization (Shi et al. 2020)

7.3 Microfluidics

Global well-being might be significantly improved by giving poor people access to healthcare, particularly at the time of the COVID-19 epidemic. Numerous diagnostic tools, however, are prohibitively expensive and difficult to use in the field under exceptionally tough circumstances when advanced laboratory facilities and qualified personnel are hard to come by. Innovative techniques for diagnosis that utilize microfluidics are being successfully created and are now being employed in rural healthcare institutions to suit the needs of the underserved population.

Fluid dynamics and transport of nutrients inside process equipment can be more thoroughly described using CFD technology. Combining experimental approaches (like laser Doppler anemometry (LDA) and particle-image velocimetry (PIV)) alongside CFD modelling allows for the characterization of 3D flow fields within a microfluidic chip. The accompanying rates of flow and pattern can be theoretically examined and measured prior to the creation of a microfluidic chip. Specific variables such as velocity of fluid intake and shear stress may be modified as well to better predictions and hence improve the design and experiment. Whenever it is impractical to install probes inside fluid domains to monitor values like pressure and velocity, CFD algorithms enable visualization of flow events.

Because CFD programs are highly scalable, multiple processors can work in parallel to tackle the assigned complex task in less time. Model meshing is done prior to solutions and must be sufficiently developed to collect data at specific points that are important within a microfluidic chip. As a result, grid independence checks must be performed prior to running the simulations. Navier–Stokes equations characterizing flow inside the microfluidic chip at different rates of flow can be solved using commercially available software tools. The geometrical arrangement and structure of the microstructures-oriented flow preferentially in a particular direction, assisting in the construction of a three-dimensional flow. The use of both CFD modelling and experimental evaluation can result in a well-designed 3D study design (Fig. 4).

Biomechanical factors, such as fluid shear stresses, have a significant influence during the research of cell activity in vitro. The surface properties of the substrate, microstructure, and surface conditions all influence shear force development, which

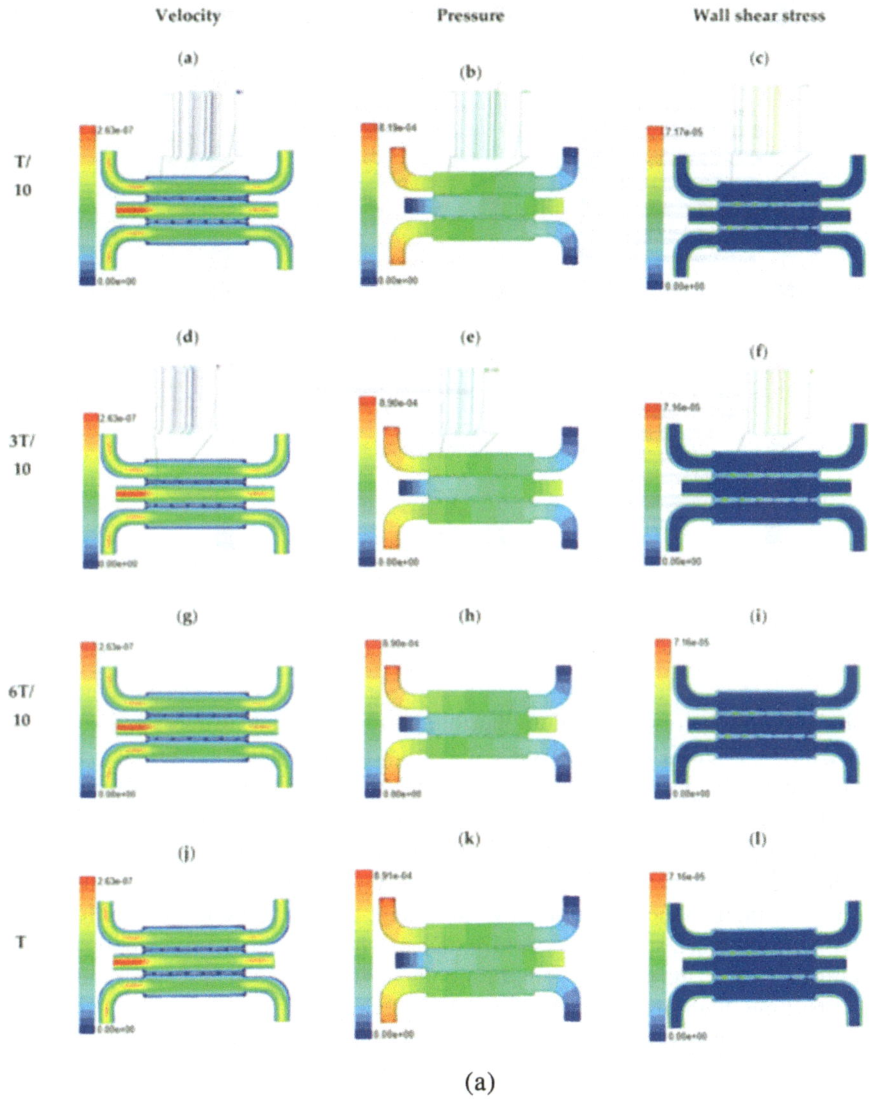

Fig. 4 Contours of velocity, pressure and wall shear stress inside different microfluidic channels: **a** multilayered and **b** sandwich model for studying wound healing at different flow rates (Gupta et al., 2022)

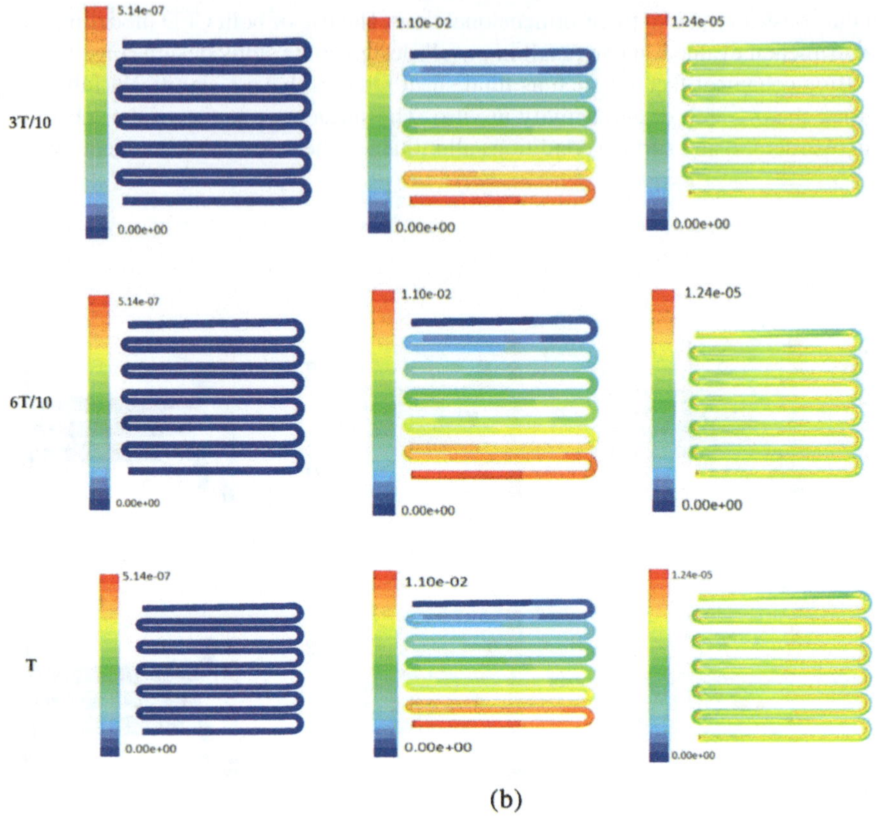

Fig. 4 (continued)

influences cellular responses like adhesion, proliferation, and differentiation (Huang et al., 2010). Even CFD can provide a scientific simulation that can produce spatially and time-resolved predictions for a number of respiratory drug delivery-related difficulties, ranging from early aerosol formation through respiratory cellular drug absorption (Longest et al. 2019). Three-dimensional computational fluid dynamics models have been developed to more precisely represent the impacts of anatomic detail and their influence on inhaled substance transport (Longest and Holbrook, 2012). The human respiratory system, which extends through the conducting airways within the lung to the external nares or mouth, has been studied in several studies. For the first time, high-resolution MRI and X-ray computed tomography (CT) imaging have been employed to give anatomic detail for these models, which were used to assess specific to the location airflows and local tissue dosimetry in both the lower and upper respiratory systems of people. Researchers coupled these observations with chemical-specific boundary conditions. The difficulty in acquiring high-resolution (HR) data on three-dimensional geometries and the lack of computer software and technological resources to solve these challenges have been the primary obstacles to constructing thorough CFD models that take into consideration all airway areas.

7.4 Drug Administration

The most recent developments in CFD applications for studying drug distribution are presented in this section (Fig. 5). Drugs for common issues include nasal congestion and allergic rhinitis which are delivered via nasal sprays. In addition, it is utilized as an alternative to injections and tablets for systemic treatment. However, it is a complex physical process that is dependent on a number of elements, including the architecture of the nasal device, the liquid properties of the medication, and how the patient is handled. Because inhaled air combines with the intended pharmaceutical formulation, an additional phase (such as a solid or liquid) is introduced. As a result, nasal medication administration research can naturally be a multiphase flow application incorporating liquid–gas (liquid jet flow through the nasal air cavity space amid nasal irrigation), solid–gas (drug particle delivered to the nasal air cavity upon nasal spray atomization), and even gas–gas (gas vapour combined with inhaled air during vaporized drug formulations) (Shrestha et al. 2021).

The capacity for predicting the distribution of spatial and temporal concentrations of drug distribution in the eyes is required for the quantitative investigation of the therapeutic effect and overdose issues via various topical delivery strategies. To achieve these goals and investigate the effect of delivery rate and interval on the concentration distributions, researchers developed a virtual human eye model with several physiologically accurate ocular compartments that utilized CFD. For a numerical examination of how the delivery approach may alter the transport of drugs and concentration distribution in the human eye over time (Yi et al. 2022).

Similarly, by using patient-specific boundary conditions, CFD can aid in obtaining a more exact distribution of drugs estimation in trans-arterial embolization management. For a minimally invasive therapy for advanced liver cancer, microspheres containing a chemotherapeutic drug or radioactive yttrium-90 (90Y) are introduced into the hepatic artery tree by catheter. The injection dose and location should be carefully chosen to optimize microsphere distribution in the liver for customized treatment. CFD simulations of the flow of blood in the hepatic artery, when properly parameterized, can aid in approximating this distribution (Taebi et al., 2020).

7.5 Simulating Microswimmers

Microswimmers are seen as potential agents in biological applications like targeted medication administration, less invasive surgery, and cell manipulation. Based on observations of bacteria and spermatozoa, their propulsion principles are considered feasible actuation methods for microswimmers. Artificial microswimmers offer

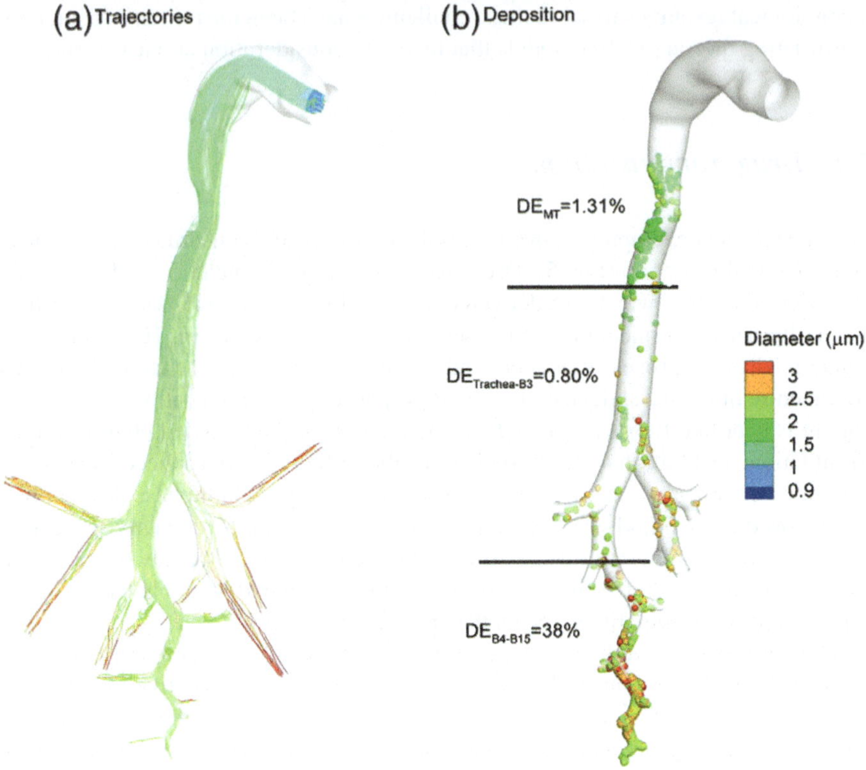

Fig. 5 CFD simulation showing delivery of an originally aerosol by enhanced condensational growth (ECG) with a humidity stream temperature of 39 °C in terms of **a** trajectories and **b** deposition locations (Longest et al., 2012)

enormous potential in a wide range of microfluidic and therapeutic applications. The precise estimation of the 3-D trajectory of micro/nanoswimmers is a critical component in achieving high-precision control of motion in therapeutic applications. Viscous forces dominate rigid-body kinematics in robotic systems. To investigate the generated flow field which surrounds a swimmer, a verified CFD simulation needs to be used. Parameters such as 3D trajectories, propulsion, and tangential velocities can be the outcomes of modelling used to validate the model. To model the motion of swimmers resembling bacteria in a viscous medium, force-free swimming restrictions are used. A study understanding the technique employed to estimate the fluid restriction that the body of a swimmer encounters is required. According to parametric studies, the hydrodynamic relationship that exists between the body and the tail is critical for anticipating the trajectory predictions of such systems. A rotating magnetic field can be utilized to remotely control magnetized swimmers with helical tails in a biocompatible manner, eliminating the need for onboard propulsion machinery. CFDs can help us understand the actions of microswimmers. The swimmer's velocity can be examined with induced flow fields, and hydrodynamic forces operating upon the swimmer's surface can be estimated using the CFD model.

It's crucial to comprehend the swimmers' trajectories for their controlled swimming. In confined spaces, two unique trajectories stand out: one is a straight route near the channel centre when the tail pulls the head, and the other is a helical trajectory close to the channel limits when the tail pushes the head. The step-out behaviour, in which the magnetically actuated swimmer loses synchronization with the spinning magnetic field due to excessive viscous torque, is another element influencing the trajectories of artificial swimmers (Caldaget al., 2018).

The research explains hydrogel-based, magnetically driven and controlled, degradable microswimmer for the delivery and release of theranostic cargo that is sensitive to the pathological indicators in its microenvironment (Ceylan et al., 2019). They considered microswimmer to be contained within an incompressible fluid domain with no-slip boundary conditions. The whole microswimmer surface achieves a unique velocity under the required swimming circumstances.

Tabak et al. used the RFT model in conjunction with complex-impedance research to take into consideration flow field interaction and precisely predict instantaneous fluid resistance under viscous swimming conditions. In addition to investigating swimming velocities using induced flow fields, the authors used the CFD model to calculate the hydrodynamic forces occurring on the swimmer surface (Tabak et al. 2013).

Microswimmers maintain helical or linear trajectories based on whether or not the tail is pushing or pulling the swimmer. The scientists used a direct computational technique which integrates a basic kinematics model with a CFD simulation to resolve the Stokes equation for swimming with low Reynolds numbers in order to better understand this behaviour. The kinematic model changes the swimmer's position and orientation at every point via the CFD model's linear and angular velocities. In addition to the viscous force, the CFD model considers the swimmer's external magnetic torque, gravity force, and normal contact force. Their findings not only deepen our grasp of the fundamentals, but they also have significant implications for the creation of microswimming robot control systems (Caldag et al. 2019).

7.6 CFD in Cardiovascular Applications

The use of CFD is intended to theoretically greatly benefit cardiovascular medicine, and clinical trials, improve diagnostic evaluation, and device design to forecast physiological reactions to intervention, and calculate the preceding hemodynamics parameters that are not measurable (Fig. 6). The three primary physiologies of the heart's functions—valves, arteries, and ventricles—are addressed in research on CFD applications concerning the cardiovascular system. The term "cardiovascular" refers to heart disease, which is the leading cause of mortality worldwide. Aortic valve constriction of blood flow leakage on the valve leaflet is the two main causes of heart valve disease (Pedrizzetti et al. 2003). The researchers used a mix of MRI and CFD

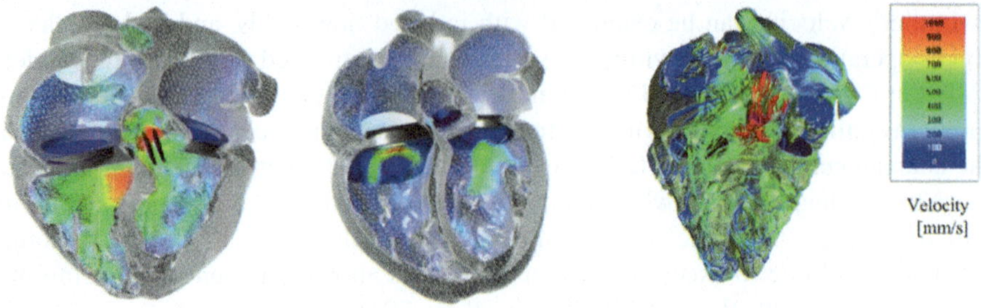

Fig. 6 Velocity profile for human heart blood flow calculated using numerical modelling and simulations (Aksenov et al., 2020)

modelling to investigate the hemodynamic consequences of different valve openings. The authors studied the hemodynamic parameters of velocity, pressure, and wall shear stress to evaluate the blood response to severe aortic stenosis. The result shows a significant drop in the blood pressure in the tiny valve opening, resulting in a blockage in the blood ejection due to the narrowing of the valve. As a result of the findings, the lower leaflet aperture had an effect on the blood flow and enhanced leaflet stress.

7.7 *Fluid–Structure Interaction to Investigate the Upper Airway of Obstructive Sleep Apnoea (OSA) Patients*

OSA is a respiratory disorder caused by upper airway blockage during sleep. The nasal valve region is regarded to be the most restricted and collapsible within the whole nasal cavity. The velopharynx is regarded to be a highly collapsible region of the nasal cavity and the flow-limiting component in persons with OSA during sleeping (Jeong et al., 2007). Continuous positive airway pressure (CPAP) therapy is considered the gold standard, but it cannot always be used due to issues with operability, mobility, or a sense of continual incongruity with the mask. Cephalograms, CT, and MRI and endoscopy are examples of clinical procedures used for determining the precise location of the airway blockage. These techniques, however, are unable to determine the condition of upper airway ventilation since they only provide information in two dimensions in a minimally sophisticated three-dimensional (3D) format. CFD can analyse airflow in a manner comparable to that observed during actual breathing, especially in instances involving upper airways with complex geometry, making it extremely useful for assessing upper-airway ventilation conditions. 3D reconstruction of the upper airway was produced from CT data using volume-rendering software. Mesh convergence analysis must be carried out on models with varying grid scales. To examine the ventilation situation, researchers are now adopting CFD research, which reconstructs airflow using a 3D upper airway model (Lu et al., 2012).

7.8 Fluid Dynamics in Syringomyelia Cavities

This study used serial anatomical MR and PC-flow images data from a patient with Chiari I malformation and syringomyelia. The purpose of this study was to determine the spectrum of syrinx fluid velocities during various physiological conditions (Fig. 7). To investigate syrinx and cerebrospinal fluid (CSF) velocities, the rate of flow of syrinx fluid was computationally predicted. They also looked at how the heart rate, CSF velocity waveform, CSF velocity, and syrinx size affected the flow of syrinx fluid in the model. They concluded that fluid within the syrinx cavities moves during the cardiac cycle, affected by CSF flow and heart rate (Vinje et al., 2018).

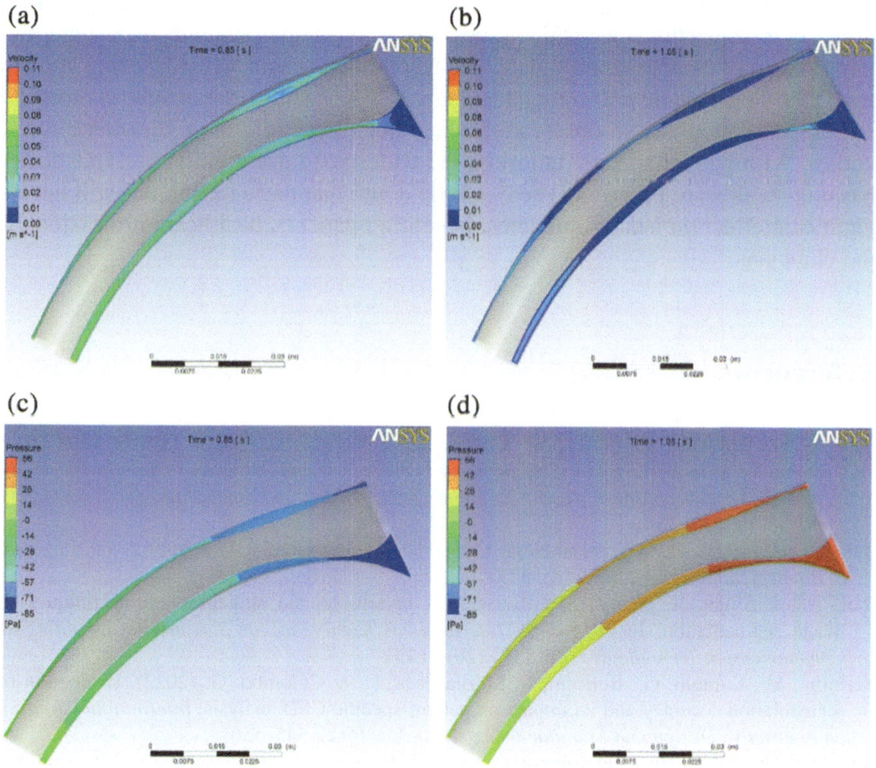

Fig. 7 Computational fluid dynamic modelling of cerebrospinal fluid pressure in Chiari malformation and syringomyelia. CSF pressures at two-time steps from a simulation are shown in (**a**-**b**) and (**c**-**d**), respectively, in representative model outputs (Clarke et al., 2013)

8 Conclusion

CFD in biomedical applications is still in its early stages due to the intricacy of human physiology and the fluid nature of the human body. However, the development of digital computers with high-speed hardware and software has made it more affordable and practical. The advancement of biomedical approaches and methods in CFD has been sparked by the rising importance of understanding how body fluids and system components function and the biofluid physiology study during the past several years. Regulatory authorities have acknowledged the outstanding substantial computational modelling and technology difficulties that have emerged quickly and are geared towards the development of CFD. The capabilities of CFD simulation tools, excellent creative models, and the development of fresh applications for simulating complicated fluid mechanics issues concerning human anatomy of the hemodynamics, cardiovascular, neurological, and respiratory systems are now being applied gradually. Consequently, it's crucial to show how well simulation results compare to invasive measurements in observational research, especially in multicenter clinical studies. With patient-specific tailoring, the widespread use of CFD will drastically speed up healthcare progress. There is little doubt that these techniques will have a significant effect on health outcomes, benefiting patients, healthcare professionals, and clinicians.

References

Aksenov, A., Zhluktov, S., Zietak, W., Cotton, R., & Vučinić, D. (2020). Human heart blood flow numerical modelling and simulations. *Advances in Visualization and Optimization Techniques for Multidisciplinary Research: Trends in Modelling and Simulations for Engineering Applications*, 237–263.

Baieth, H. A. (2008). Physical parameters of blood as a non-Newtonian fluid. *International Journal of Biomedical Science: IJBS, 4*(4), 323.

Bali, A., & Singh, S. N. (2015, February). A review on the strategies and techniques of image segmentation. In *2015 Fifth International Conference on Advanced Computing & Communication Technologies* (pp. 113–120). IEEE.

Bertolini, M., Luraghi, G., Belicchi, I., Migliavacca, F., & Colombo, G. (2022). Evaluation of segmentation accuracy and its impact on patient-specific CFD analysis. *International Journal on Interactive Design and Manufacturing (IJIDeM), 16*(2), 545–556.

Bluestein, D. (2017). Utilizing computational fluid dynamics in cardiovascular engineering and medicine—What you need to know. *Its translation to the clinic/bedside. Artificial organs, 41*(2), 117.

Brindise, M. C., Rothenberger, S., Dickerhoff, B., Schnell, S., Markl, M., Saloner, D., Rayz, V. L., & Vlachos, P. P. (2019). Multi-modality cerebral aneurysm haemodynamic analysis: in vivo 4D flow MRI, in vitro volumetric particle velocimetry, and in silico computational fluid dynamics. *Journal of the Royal Society Interface, 16*(158), 20190465.

Caldag, H. O., & Yesilyurt, S. (2018, June). Dynamics of artificial helical microswimmers under confinement. In *International Conference on Nanochannels, Microchannels, and Minichannels* (Vol. 51197, p. V001T13A001). American Society of Mechanical Engineers.

Caldag, H. O., & Yesilyurt, S. (2019). Trajectories of magnetically-actuated helical swimmers in cylindrical channels at low Reynolds numbers. *Journal of Fluids and Structures, 90*, 164–176.

Castro, M. A., Putman, C. M., & Cebral, J. R. (2006). Computational fluid dynamics modelling of intracranial aneurysms: Effects of parent artery segmentation on intra-aneurysmal hemodynamics. *American Journal of Neuroradiology, 27*(8), 1703–1709.

Ceylan, H., Yasa, I. C., Yasa, O., Tabak, A. F., Giltinan, J., & Sitti, M. (2019). 3D-printed biodegradable microswimmer for theranostic cargo delivery and release. *ACS Nano, 13*(3), 3353–3362.

Clarke, E. C., Fletcher, D. F., Stoodley, M. A., & Bilston, L. E. (2013). Computational fluid dynamics modelling of cerebrospinal fluid pressure in Chiari malformation and syringomyelia. *Journal of Biomechanics, 46*(11), 1801–1809.

Dar, A. S., & Padha, D. (2019). Medical image segmentation: A review of recent techniques, advancements and a comprehensive comparison. *International Journal of Computational Science and Engineering, 7*(7), 114–124.

Dhankhar, P., & Sahu, N. (2013). A review and research of edge detection techniques for image segmentation. *International Journal of Computer Science and Mobile Computing, 2*(7), 86–92.

Dyedov, V., Einstein, D. R., Jiao, X., Kuprat, A. P., Carson, J. P., & Del Pin, F. (2009). Variational generation of prismatic boundary-layer meshes for biomedical computing. *International Journal for Numerical Methods in Engineering, 79*(8), 907–945.

Gould, S., Gao, T., & Koller, D. (2009). Region-based segmentation and object detection. *Advances in neural information processing systems, 22.*

Guo, Z., Li, X., Huang, H., Guo, N., & Li, Q. (2019). Deep learning-based image segmentation on multimodal medical imaging. *IEEE Transactions on Radiation and Plasma Medical Sciences, 3*(2), 162–169.

Gupta, S., Patel, L., Mitra, K., & Bit, A. (2022). Fibroblast derived skin wound healing modelling on chip under the influence of micro-capillary shear stress. *Micromachines, 13*(2), 305.

He, Y., Northrup, H., Le, H., Cheung, A. K., Berceli, S. A., & Shiu, Y. T. (2022). Medical image-based computational fluid dynamics and fluid-structure interaction analysis in vascular diseases. *Frontiers in Bioengineering and Biotechnology, 10*, 671.

Huang, M., Fan, S., Xing, W., & Liu, C. (2010). Microfluidic cell culture system studies and computational fluid dynamics. *Mathematical and Computer Modelling, 52*(11–12), 2036–2042.

Inthavong, K., Wen, J., Tu, J., & Tian, Z. (2009). From CT scans to CFD modelling–fluid and heat transfer in a realistic human nasal cavity. *Engineering Applications of Computational Fluid Mechanics, 3*(3), 321–335.

Ishida, F., Tsuji, M., Tanioka, S., Tanaka, K., Yoshimura, S., & Suzuki, H. (2021). Computational fluid dynamics for cerebral aneurysms in clinical settings. *Trends in Cerebrovascular Surgery and Interventions*, 27–32.

Jamari, J., Ammarullah, M. I., Santoso, G., Sugiharto, S., Supriyono, T., & van der Heide, E. (2022). In silico contact pressure of metal-on-metal total hip implant with different materials subjected to gait loading. *Metals, 12*(8), 1241.

Jeong, S. J., Kim, W. S., & Sung, S. J. (2007). Numerical investigation on the flow characteristics and aerodynamic force of the upper airway of patient with obstructive sleep apnea using computational fluid dynamics. *Medical Engineering & Physics, 29*(6), 637–651.

Johnston, W. W., & Elson, C. E. (2008). Respiratory tract. In *Comprehensive Cytopathology* (pp. 303–371). WB Saunders.

Lee, B. K. (2011). Computational fluid dynamics in cardiovascular disease. *Korean Circulation Journal, 41*(8), 423–430.

Lim, S. H., Adib, M. A. H. M., Abdullah, M. S., Taib, N. H. M., Hassan, R., & Abd Aziz, A. (2020). Study of extracted geometry effect on patient-specific cerebral aneurysm model with different threshold coefficient (Cthres). *CFD Letters, 12*(10), 1–14.

Lintermann, A. (2021). Computational meshing for cfd simulations. Clinical and biomedical engineering in the human nose: A computational fluid dynamics approach, 85–115.

Longest, P. W., & Holbrook, L. T. (2012). In silico models of aerosol delivery to the respiratory tract—Development and applications. *Advanced Drug Delivery Reviews, 64*(4), 296–311.

Longest, P. W., Tian, G., Walenga, R. L., & Hindle, M. (2012). Comparing MDI and DPI aerosol deposition using in vitro experiments and a new stochastic individual path (SIP) model of the conducting airways. *Pharmaceutical Research, 29*, 1670–1688.

Longest, P. W., Bass, K., Dutta, R., Rani, V., Thomas, M. L., El-Achwah, A., & Hindle, M. (2019). Use of computational fluid dynamics deposition modeling in respiratory drug delivery. *Expert opinion on drug delivery, 16*(1), 7–26.

Lu, M. Z., Liu, Y., & Ye, J. Y. (2012). Turbulent flow pattern in upper airway with obstructed sleep apnea. In *15th International Symposium on Flow Visualization*, ISFV15.

Minaee, S., Boykov, Y. Y., Porikli, F., Plaza, A. J., Kehtarnavaz, N., & Terzopoulos, D. (2021). Image segmentation using deep learning: A survey. IEEE transactions on pattern analysis and machine intelligence.

Narendrakumar, L., & Ray, A. (2022). Respiratory tract microbiome and pneumonia. *Progress in Molecular Biology and Translational Science, 192*(1), 97–124.

Niu, Z., & Li, H. (2019, June). Research and analysis of threshold segmentation algorithms in image processing. In *Journal of Physics: Conference Series* (Vol. 1237, No. 2, p. 022122). IOP Publishing.

Pandey, R., Kumar, M., & Srivastav, V. K. (2020). Numerical computation of blood hemodynamic through constricted human left coronary artery: Pulsatile simulations. *Computer Methods and Programs in Biomedicine, 197*, 105661.

Pedrizzetti, G., & Perktold, K. (Eds.). (2003). *Cardiovascular fluid mechanics*. Springer.

Prabowo, A. R., Sohn, J. M., & Triyono, J. (2020). Finite element analysis of different artificial hip stem designs based on fenestration under static loading. *Procedia Structural Integrity, 27*, 101–108.

Reid, L. (2021). An introduction to biomedical computational fluid dynamics. *Biomedical Visualisation, 10*, 205–222.

Richmond, C. (2004). Sir godfrey hounsfield.

Sharp, J. C., & King, S. B. (2010). MRI using radiofrequency magnetic field phase gradients. *Magnetic Resonance in Medicine: An Official Journal of the International Society for Magnetic Resonance in Medicine, 63*(1), 151–161.

Shi, B., & Huang, H. (2020). Computational technology for nasal cartilage-related clinical research and application. *International Journal of Oral Science, 12*(1), 21.

Shishir, S. S., Miah, M. A. K., Islam, A. S., & Hasan, A. T. (2015). Blood flow dynamics in cerebral aneurysm-A CFD simulation. *Procedia Engineering, 105*, 919–927.

Shrestha, K., Salati, H., Fletcher, D., Singh, N., & Inthavong, K. (2021). Effects of head tilt on squeeze-bottle nasal irrigation–A computational fluid dynamics study. *Journal of Biomechanics, 123*, 110490.

Sotiropoulos, F. (2012). Computational fluid dynamics for medical device design and evaluation: Are we there yet? *Cardiovascular Engineering and Technology, 3*, 137–138.

Srivastav, V. K., Paul, A. R., & Jain, A. (2011). Computational Fluid Dynamics Study of Airflow and Particle Transport in Third to Sixth Generation Human Respiratory Tract. *International Journal of Emerging Multidisciplinary Fluid Sciences, 3*(4).

Steinman, D. A. (2002). Image-based computational fluid dynamics modelling in realistic arterial geometries. *Annals of Biomedical Engineering, 30*, 483–497.

Steinman, D. A., Milner, J. S., Norley, C. J., Lownie, S. P., & Holdsworth, D. W. (2003). Image-based computational simulation of flow dynamics in a giant intracranial aneurysm. *American Journal of Neuroradiology, 24*(4), 559–566.

Tabak, A. F., & Yesilyurt, S. (2013). Improved kinematic models for two-link helical micro/nano swimmers. *IEEE Transactions on Robotics, 30*(1), 14–25.

Taebi, A., Pillai, R. M., S. Roudsari, B., Vu, C. T., & Roncali, E. (2020). Computational modelling of the liver arterial blood flow for microsphere therapy: Effect of boundary conditions. *Bioengineering, 7*(3), 64.

Temor, L., Cancelliere, N. M., MacDonald, D. E., Coppin, P. W., Pereira, V. M., & Steinman, D. A. (2022). Integrating computational fluid dynamics data into medical image visualization workflows via DICOM. *International Journal of Computer Assisted Radiology and Surgery, 17*(6), 1143–1154.

Tsega, E. G. (2018). Computational fluid dynamics modelling of respiratory airflow in tracheobronchial airways of infant, child, and adult. *Computational and Mathematical Methods in Medicine.*

Uchiyama, Y., Fujimura, S., Takao, H., Suzuki, T., Hayakawa, M., Ishibashi, T., Karagiozov, K., Fukudome, K., Murayama, Y., & Yamamoto, M (2021). Hemodynamic investigation of the effectiveness of a two overlapping flow diverter configuration for cerebral aneurysm treatment. *Bioengineering, 8*(10), 143.

Vinje, V., Brucker, J., Rognes, M. E., Mardal, K. A., & Haughton, V. (2018). Fluid dynamics in syringomyelia cavities: Effects of heart rate, CSF velocity, CSF velocity waveform, and craniovertebral decompression. *The Neuroradiology Journal, 31*(5), 482–489.

Weibel, E. R., & Gomez, D. M. (1962). Architecture of the human lung: Use of quantitative methods establishes fundamental relations between size and number of lung structures. *Science, 137*(3530), 577–585.

Yang, J., Ni, S., & Weng, W. (2017). Modelling heat transfer and physiological responses of unclothed human body in hot environment by coupling CFD simulation with thermal model. *International Journal of Thermal Sciences, 120*, 437–445.

Yi, H., Feng, Y., & Gappa-Fahlenkamp, H. (2022). Analysis of topical dosing and administration effects on ocular drug delivery in a human eyeball model using computational fluid dynamics. *Computers in Biology and Medicine, 141*, 105016.

Zamir, M., Moore, J. E., Fujioka, H., & Gaver, D. P. (2010). Biofluid mechanics of special organs and the issue of system control: sixth international bio-fluid mechanics symposium and workshop, March 28–30, 2008 Pasadena, California. *Annals of Biomedical Engineering, 38*, 1204–1215.

Zhong, L., Zhang, J. M., Su, B., Tan, R. S., Allen, J. C., & Kassab, G. S. (2018). Application of patient-specific computational fluid dynamics in coronary and intra-cardiac flow simulations: Challenges and opportunities. *Frontiers in Physiology, 9*, 742.

Chapter 5
Computational Fluid Dynamics in the Human Integumentary Systems

1 Introduction

The human skin can be simulated using computational fluid dynamics (CFD) to provide detailed information that cannot be obtained from experimentation. Understanding the geometric complexity of the skin anatomy, grid generation, and boundary conditions are all part of the CFD study. Several topics from CFD investigations are presented in this book chapter, including stress and strain, diffusion, wound healing, and microneedle insertion analysis. Simple geometry can calculate these parameters with minimal processing power, while more humanistic geometry can yield more precise results.

2 Background

The human skin also known as the integumentary system serves as the body's outward defence (McLafferty et al., 2012). The components of the integumentary system are the skin, epidermal derivatives, and the skin's sensory systems (Rehfeld et al., 2017). In terms of surface area, skin is indeed the largest organ of the human body, making up roughly 16% of an adult's total body weight. Human skin shields the body against chemical, biological, mechanical, and thermal impacts. It is a crucial component that protects internal organs against thermal shock, microbial infection, UV radiation, and mechanical harm. This makes it extremely prone to harm, which would have a substantial effect on both individuals and patients as well as the healthcare industry. The epidermis and dermis are the two primary layers of the skin. The epidermis is the outermost layer of the skin, followed by the dermis and hypodermis. The epidermis is the most superficial layer of the skin, where the epithelium forms a barrier. Dermis is a layer deep to the epidermis that contains connective tissue that gives the skin strength and elasticity (Casale et al., 2016; Ponmozhi et al., 2021). The hypodermis

S. E. Jujjavarapu et al., *Computational Fluid Dynamics Applications in Bio and Biomedical Processes*, https://doi.org/10.1007/978-981-99-7129-9_5

has historically been referred to as part of the skin since it connects the surface to deeper structures. Each layer has a distinct structure and function that contributes to the overall structural function. Hair, hair follicles, nails, and glands are examples of epidermal derivatives. Also, each layer has resident immune cells that are constantly scanning the skin for injury in addition to these different cell types. The complex synchronization of numerous distinct cell types in orderly steps is necessary for skin restoration (Henrotet al., 2020).

Since the importance of understanding how bodily fluids and system components work and biofluid physiology research has grown in recent years, biomedical techniques and technology have advanced (Reid, 2021). Biomedical research using CFD software is still in its early stages, with modelling incorporating the physiology and pathophysiology of the integumentary system. For medical and cosmetics studies, such as the creation of skincare products, the understanding of skin diseases, and the impact of chirurgical or pharmacological acts, it is important to examine the responses of different skin layers. The fluids and body structures in the human body must interact in a complex way for the body to function effectively. Fluids are a component of all biological systems, but the integumentary system due to its function in controlling perspiration production, burn injuries, heat distribution, and arterial blood temperature, is the most obviously impacted. The skin is the primary tissue that balances the fluid imbalance in the human body. Consequently, it appears to be quite intriguing to examine how fluids diffuse and are distributed through human skin. Also, knowing and accurately anticipating the biomechanical properties of human skin are critical for medical applications. The planning of reconstructive surgery and wound healing depends on deformations and stress distribution within the skin. Precise models of the skin's response under various mechanical loading circumstances are required to investigate and reveal the mechanical and mechanobiological processes (Ponmozhi et al., 2021).

By providing opportunities for simulation before making a real commitment to create treatment options in the right direction and to carry out any medical design adjustment, CFD plays a significant role. Due to the significance of computational medical models, research into biomedical CFD applications has attracted a lot of attention recently. CFD models are constantly being translated into therapeutic tools that doctors can use to treat a range of different skin disorders. Thus, with a focus on skin-related applications, this chapter examines the CFD study employing the most recent developments in the clinical field. The next section discusses the use of CFD aimed at theoretically extremely beneficial areas such as drug delivery, clinical trials, improved diagnostic assessment, and device design to predict physiological reactions to intervention and compute previously unmeasured parameters. Many simulation and clinical outcomes studies, including toxicology studies, layer-specific analysis of mechanical behaviour, diffusion studies, and shear stress assessments on the skin, have been conducted. Most studies considered skin as multilayered while performing the computational analysis. The associated methodological, and analytical assessment, and a description of three key physiologies of skin, namely epidermis, dermis, and hypodermis, have been addressed.

3 Anatomy and Physiology of the Integumentary System

The skin is an extremely complex biological system with a plethora of connected physical processes that function in concert or sequentially. This can be seen in the case of the healing process, where an injury initiates a cascade of biochemical and mechanobiological activities. From the standpoint of mechanical and material science, the skin is a multiphasic and multiscale structure with a diverse biological nature. The skin is made up of a thick outermost layer, a large layer of fatty tissue beneath the skin's surface, and a system of sweat glands that are widely dispersed and temperature-sensitive. A large number of touch-sensitive, pressure-sensitive, and temperature-sensitive cells are also present in the skin. The skin is a complex array of tissues that serve a variety of distinct yet vital functions such as serving as a barrier by guarding against UV rays, mechanical wear and tear, chemical agents, microorganisms (that function as part of the innate immune system), and evaporation (which maintains the body's fluid balance). Skin aids in temperature regulation by cooling via directing blood through superficial blood vessels or sweating and staying warm by diverting blood away from superficial blood vessels (Romanovsky, 2014). Skin secretes as well since it has a range of exocrine glands, including eccrine sweat glands and sebaceous glands (Ruela et al. 2016). It is a heterogeneous multilayered tissue whose major role is to defend itself from the surrounding environment by acting as an effective barrier to exogenous molecule absorption, however allowing permeation of selective drugs through skin permeation routes. Also, the skin is the body's greatest sense organ, as its sensory nerve receptors detect a variety of stimuli, including mechanical (pressure or stretching) and thermal (heat and cold) (Richardson, 2003).

The epidermis and dermis are the primary layers of human skin, as depicted in Fig. 1. The epidermis is the skin's outermost impermeable layer of healthy skin and consists of keratinocytes, melanocytes, Langerhans cells, and Merkel cells (Yousef et al., 2022). The epidermis lacks vascularization; therefore, nutrients must diffuse from the dermo-epidermal junction to keep it alive. In the epidermis, five layers symbolize the various stages of cell life. The germinative (or bottom) layer, stratum spinosum, stratum granulosum, stratum lucidum, and stratum corneum are the layers that go from inside to outside. The sebaceous glands, sweat glands, and hair follicles are all located in the epidermis. Keratinocytes are ectodermal cells. A keratinocyte takes around four weeks to proliferate, develop, and migrate from the stratum basale to the stratum corneum, implying that the skin regenerates in four weeks. Keratinocytes migrate upwards from the basal layer and undergo differentiation throughout their lives (Piipponen et al., 2020). Hence, the study of this migration is important.

The dermis gives the skin strength, nutrition, and immunity and is abundant in extracellular matrix (ECM), vascular, and mechanoreceptors. The dermis is supported by subcutaneous adipose tissue, which serves as an energy store. Moreover, it provides the dermis with growth factors continuously. In contrast to the epidermis, the dermis has few cells and is primarily made up of collagen and elastin

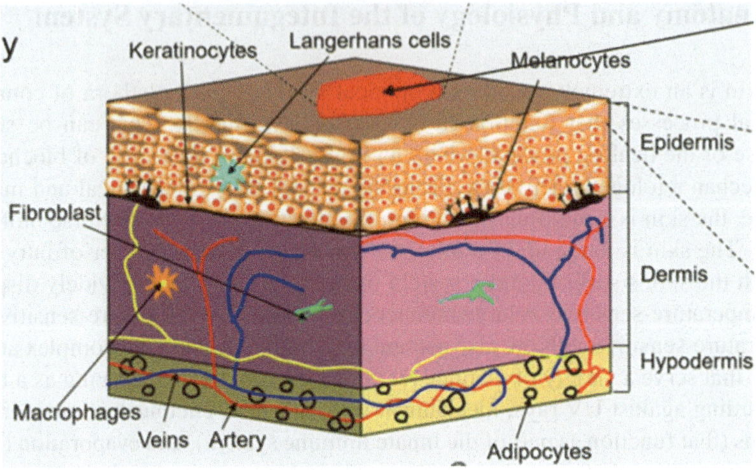

Fig. 1 Diagram of the human integumentary system

proteins. The papillary dermis is located adjacent to the dermal–epidermal junction. The reticular dermis is located beneath it and is the dermis's deeper layer.

The hypodermis is the skin's lowest layer. The hypodermis is defined by a rapid transition from fibrous dermal matrix to adipocyte-rich tissue (Fig. 1).

The discipline of computational fluid dynamics (CFD) makes it feasible to solve fluid motion equations and generate qualitative and quantitative estimations of fluid flow events that would not otherwise be achievable. To create a trustworthy CFD simulation, several factors should be taken into account and procedures should be followed. They include creating a suitable geometry that most closely mimics the primary characteristics of genuine geometry. The geometry is then discretized, which divides it into several small components that may be generated using different mesh creation techniques. Numerical methods are used in CFD to solve the conservation rules for each of those small components. There are numerous methods for creating meshes, a few of which are automated, meaning that when the geometry is given, the mesh will normally be created with a predetermined resolution. If required, a mesh can be improved (had more components added) to attain a better level of precision. Nonetheless, when employing structured mesh generation techniques on basic geometries, automated mesh production is typically effective. Complex geometries typically require a combination of structured and unstructured meshes, both of which must typically be created manually. The following stage is to decide whether boundary conditions, such as intake, solid surface, and outflow, are appropriate given the circumstances. At each time step, the CFD solver generates the pressure, shear stress, or velocity profile. Tables 1 and 2 describe the skin layer parameters that can be considered in a simulation.

Table 1 Key parameters of skin to be considered during simulation

S. No.	Parameter		Value	References
1	Thickness of stratum corneum		10–50 μm	Tayyaba et al. (2020), Mercuri et al. (2021)
2	Corneocyte	Thickness	0.8 μm	Mercuri et al. (2021)
		Diameter	40 μm	
3	Thickness of viable epidermis		80–100 μm	Tayyaba et al. (2020), Mercuri et al. (2021)
4	Thickness of dermis		1000 μm	Tayyaba et al. (2020)
5	Stratum corneum	Friction coefficient	0.42	Tayyaba et al. (2020)
		Stress failure (Mpa)	13–44	
		Density (kg/m^3)	1300	
6	Viable dermis	Friction coefficient	0.42	Tayyaba et al. (2020)
		Stress failure (Mpa)	9.1	
		Density (kg/m^3)	1250	
7	Epidermis	Friction coefficient	0.42	Tayyaba et al. (2020)
		Stress failure (Mpa)	7.3	
		Density (kg/m^3)	1200	

Table 2 Table showing thermophysical properties of different skin layers (Kandala et al. 2013)

Properties ↓	Epidermis	Papillary dermis	Reticular dermis	Fat	Muscle
Thickness (mm)	0.1	3589	1200	0.235	0
Specific heat (J/kg K)	0.8	3300	1200	0.445	0.18
Density (kg/m^3)	0.8	3300	1200	0.445	1.26
Thermal conductivity (W/m K)	2	2674	1000	0.185	0.08
Perfusion rate (10^{-3}) (1/s)	8	3600	1085	0.51	2.7
Metabolic heat rate (W/m^3)	–	3852	1030	0.558	6.3

4 Mathematical Tools for Skin CFD

Based on conservation laws, CFD employs mathematical tools to offer a quantitative forecast of fluid flow, heat transfer, or mass transfer processes. Studies involving these skin-related phenomena in humans can all benefit from their use. Mathematical models, or mathematical equations, can be used to depict and solve physical processes including fluid percolation, heat transport, and mass transfer across the dermal layers and help to solve the problem using proper CFD tools. The approach specifically entails the following steps: (1) defining the geometry of the layers of skin or the apparatuses containing tissues or artificial membranes; (2) meshing the geometry; (3) selecting the appropriate equation governing the physical phenomena; (4) solving the governing equations with appropriate numerical methods; and (5) post-processing and analysing the obtained numerical data.

4.1 Diffusion Model

The deepest layer, the dermis, includes blood capillaries that extend into deeper subcutaneous tissue. The drug diffuses across the intercellular route of the stratum corneum and epidermis before reaching the blood capillaries and being absorbed into the circulatory system.

The following factors influence drug diffusion across the skin layers: (a) Drug diffusion coefficient; (b) layer porosity; (c) time period; and (d) fluid velocity. Considering unsteady diffusion and the porosity of the cell layers, researchers formulated equations regulating drug diffusion across the skin layers that are more suited for drug diffusion experiments in microfluidic skin-on-a-chip devices (Narasimhan et al. 2015). This simple model can be utilized while considering transdermal drug delivery without suffering much from accuracy loss because it doesn't include a lot of parameters. They developed the following equations to regulate drug diffusion via the skin layers:

1. The following equation describes drug diffusion via the permeable **stratum corneum**:

$$\frac{\varepsilon^{\partial C}}{\partial t} + \nabla(Cv) = \nabla.(D_i \nabla C) \quad (1)$$

where

$$D = \varepsilon r D_f \tag{2}$$

where C = concentration of drug, v = virtual velocity field in direction perpendicular to surface of skin, ε = porosity of skin layers, r = tortuosity factor of porous medium, D_f = diffusivity, D_i = direction-dependent diffusivity of drug in stratum corneum, D = effective diffusivity.

2. The diffusion governing equation in the **porous epidermis** is given as:

$$\varepsilon \frac{\partial c}{\partial t} = \nabla.(D_i \nabla C) \tag{3}$$

3. Collagen, elastic tissue, nerve endings, hair follicles, sweat glands, and vasculature make up the fibrous structure known as the dermis. It is also conceivable to think of it as a porous medium, in which case the equation regulating drug diffusion in the porous **dermis** may be expressed as:

$$\varepsilon \frac{\partial c}{\partial t} + \nabla(Cu) = \nabla.(D_i \nabla C) \tag{4}$$

4.2 Fluid Flow Model

The constant Darcy flow velocity of blood inside the dermis is represented by u.

$$u = -\frac{K}{\mu} \frac{\partial P}{\partial x} \tag{5}$$

where $\frac{\partial P}{\partial x}$ = pressure gradient in the flow direction, μ = dynamic viscosity of the blood, K = specific permeability or intrinsic permeability of the blood through the porous structure of the dermis, and u = macroscopic velocity of blood in the dermis.

The largest organ in the human body, the skin, gives us the mechanical cues we need to interact with the outside world. Dermal blood flow is presumed that the amount of dermal flow is the same across the body. This velocity is chosen as the macroscopic Darcy momentum equation's direct solution. A proportionality between flow rate and the applied pressure differential on steady-state unidirectional flow in a homogeneous medium is denoted is more precisely written in current notation as follows:

$$\nabla.(\rho u) = 0 \tag{6}$$

5 Case Study

5.1 Strain and Stress Study

Skin is a thin transversely isotropic membrane with a nonlinear mechanical response controlled by a collagen network (Tepole, 2017). The research involving cellular pathways for wound healing mechanisms has been studied extensively, but only limited research has been carried out related to biochemical and biomechanical standpoints. Biochemical wound healing is a complex process in which fibroblasts rebuild the dermis, the skin's load- bearing layer, after injury by collagen deposition and active contraction. Hence, understanding these pathways from a biochemical and biomechanical standpoint is essential for efficient wound care.

Prestress results from the skin's inherent tension and has been shown to significantly affect skin wrinkling and wound closure. The in vivo prestress, usually ignored in wound closure models, affects the size of the extrusions. The mechanical behaviour of the skin determines the degree of strains during wound closure. To mimic reconstructive and cosmetic surgery techniques that involve incision and undermining of the skin, excision of a cutaneous defect, and closure and suturing of the wound margins, a computational model mimic displacements and strains caused by the procedure as well as skin's inherent tension. They employed an isotropic Fung-type constitutive equation for biaxial prestress tension on two circulars (large and small) wound sizes to reproduce the skin's highly nonlinear response. The model has been used to examine how the size and shape of the excision (Fig. 2), as well as the natural tension of the skin, affect the result of the procedure (Cavicchi et al., 2009).

Researchers simulated the incision, excision, and closure of skin represented by orthotropic constitutive law using finite element models. Analysis of the size of extrusions, maximum strains, and the force required to seal wounds with variously shaped excisions was performed. Out of the four shapes (elliptical, circular, fusiform, and lazy S-plasty), elliptical shape was found to have the lowest wound closure force requirement. The size of extrusions is influenced by the in vivo prestress, which is frequently disregarded in wound closure models (Flynn, 2010). According to research, not only psychological stress can cause a delay in wound healing, but mechanical stressors caused by wounding forces also have an impact on this process. This method confirms the existence of a link between wound form and wound healing. The distribution of mechanical stresses was calculated using ANSYS Workbench to calculate principal maximum stress and equivalent strain. According to research, mechanical strains caused by the pressures created by injury might also slow down the healing process. And they found that, due to the strain in the edges, rectangular and triangular wounds tend to acquire circular shapes, whereas circular shapes change into more flattened circular profiles (Rodríguez et al., 2017).

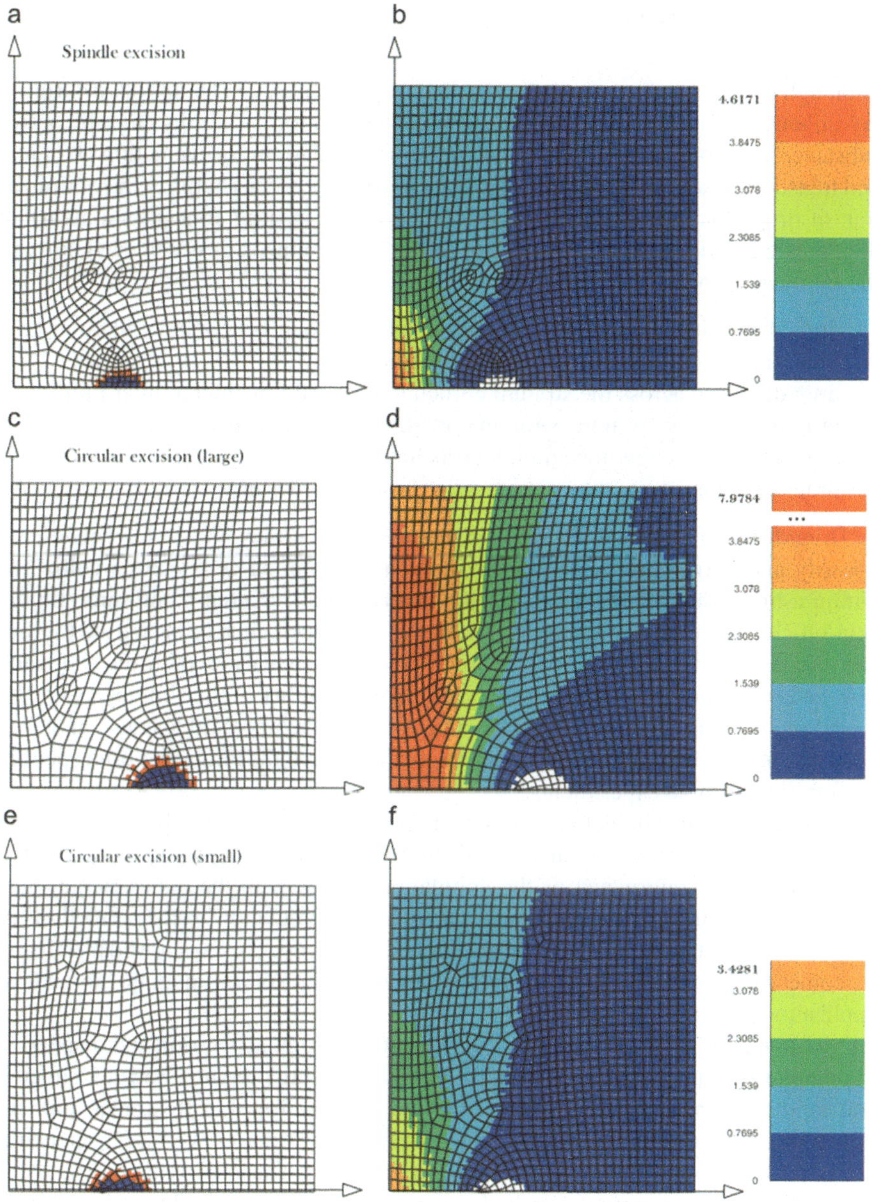

Fig. 2 Computational modelling of effects of the natural tension on skin wrinkling: **a**, **c**, and **e** shows wrinkling extension, **b**, **d**, and **f** shows the maximum Cauchy membrane forces (Cavicchi et al., 2009)

5.2 Diffusion Study

Although there are various in vitro and in vivo approaches, CFD helps in performing investigations for evaluating the cutaneous absorption of drugs and bioactive substances. The promising result calculated from simulations for evaluation of topical and transdermal administration of pharmaceutical approaches helps in the optimization of drug absorption via the skin to attain effective drug concentrations at the therapeutic location. Transport of drugs across the skin involves is achieved through the following processes:

1. drug absorption by capillary vessels, which results in systemic circulation;
2. drug dissolution and release from the formulation;
3. drug diffusion across the stratum corneum, primarily by intercellular lipids;
4. drug partitioning from the stratum corneum to viable epidermis layers;
5. drug partitioning into the stratum corneum;
6. drug diffusion across the epidermis layers into the dermis.

Drug diffusion and fluid leakage occur across the cell layers that have a definite porosity and permeability. Multiple skin layers containing cells can be cultivated within a microchannel for drug toxicity experimental research using microfluidic skin-on-chip systems. Computational fluid dynamics techniques can be used to forecast the flow and scalar transport in such devices in advance. With the predicted flow and concentration field, the porosity and permeability of the skin layers can be varied accordingly. Depending on the values of these factors, fluid leakage, and drug diffusion may or may not occur. This can greatly reduce the effort and time spent in performing multiple experiments.

Makvandi et al. simulated the skin model's static structural behaviour using ANSYS here performed an analysis of drug dispersion through the tissue model. The appropriate congruence of the simulation results with the experimental data supports the simulation methodology and additional simulation-based predictions of the behaviour of the skin model. The developed tissue model even showed promising outcomes for optical coherence tomography and ultrasound imaging biomedical applications (Makvandi et al., 2023).

Mohizin et al. studied the physics underlying microjet generation and dispersion to pinpoint crucial variables, such as filling ratio, driving pressure, nozzle diameter, and fluid type in an air-powered needle-free injector for skin treatment applications. Throughout the injection process, three separate phases were seen: penetration, reservoir development, and residual stress release. The underlying mechanisms of needle-free injectors were studied using experimental and CFD methodologies (phase and velocity profiles), and it was shown that experimental data could be reasonably predicted by the computational data, and this information may be utilized to construct and improve an air-powered needle-free injector (Fig. 3) (Mohizin et al., 2018).

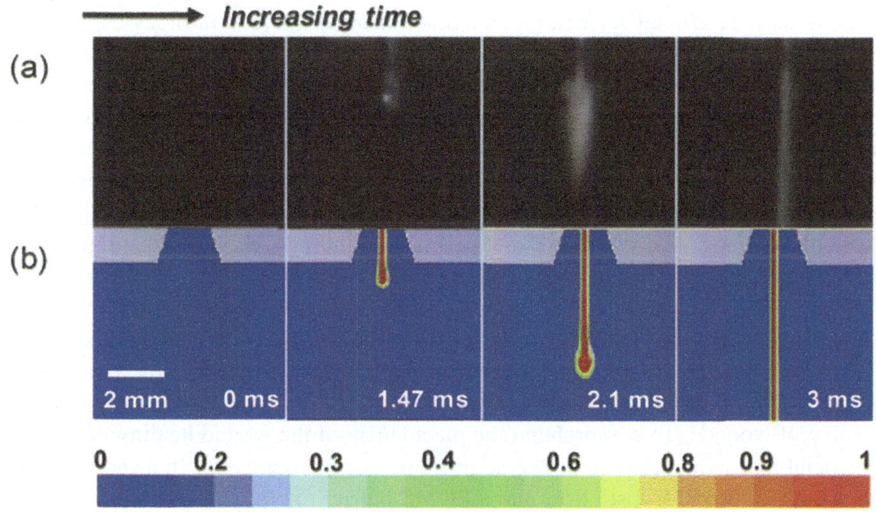

Fig. 3 A detailed representation of the initial stages of injection in a 200 m nozzle with a ratio of 0.5 v/V and driven by a force of 0.433 MPa: **a** experimental visualization **b** computational visualization (Mohizin et al., 2018)

5.3 Wound Healing Study

One of the body's most intricate processes is the healing of wounds. Several cell types with different roles during the phases of hemostasis, inflammation, proliferation, and remodelling must be coordinated in both space and time. Finite element (FE) analysis has lately been used to analyse wound closure due to ethical and biosafety concerns with human skin research. Wound healing happens on several length scales, from micro- to macrolevel remedies.

Biomechanical regulation and electric stimulation hold significant potential in the manipulation of wound healing. In clinical applications, the complexity of equipment operation and stimulation implementation continues to be a problem. Yao and co-authors developed programmable and skin temperature-activated electromechanical synergistic wound dressing made of an antibacterial electret thin film for the electric field generation and a mechanical metamaterial based on a shape memory alloy for wound contraction. They utilized ANSYS Maxwell finite element solver (AMFES) to calculate the electric field strength at the wound site to assess the effectiveness of electric field penetration. The results of the AMFES simulation demonstrated that the patterned electrode could produce an even electric field within their covered region. They concluded that the electric field speed up the migration of epithelial cells into the wound (Yao et al., 2022).

Endogenous electric fields are produced instantly when an epithelial layer is disrupted, and these fields are essential for wound healing. The orientation of the cell division axis, the direction of cell migration, and the rates of cell proliferation of cells that are crucial for healing, such as fibroblasts and keratinocytes, have all been demonstrated to be affected by electrical fields of such therapeutic intensities. Yang and co-authors described a unique galvanotaxis chamber-equipped electrical field bioreactor that is connected to a direct current power source. Electrical field simulation software ANSYS shows that the EFs in the galvanotaxis chamber are homogeneous parallel fields and show the current density distribution. Inside the galvanotaxis bioreactor, Hy926 cells were cultivated for 12 h under an online time-lapse microscope. According to the experiment's findings, Hy926 cells migrate in a clear-cut trajectory towards the cathode (Yang et al., 2008).

A microfluidic-based wound healing assay makes it simple to implement the dynamic character of the wound healing process. Many biological processes depend on fibroblast migration and proliferation, which can be controlled by a variety of microenvironmental variables. The movement of fibroblast cells is essential for the healing of wounds. To comprehend the mechanism of the wound healing process, it is crucial to conduct a quantitative analysis of this phenomenon. In a recent study, Gupta et al. utilized trypsin flow and a PDMS barrier to replicate traumatic wounds on fibroblast cell monolayers using a microfluidics system. It may be concluded that dynamic mechanotransduction phenomena of cellular entities had always developed an increased orchestra with the non-mulberry silk scaffold's biochemical niche. As a result, it encourages cell migration to the area of the wound, which speeds up wound healing. This dynamic wound healing assay allowed it easier to conduct studies on cellular pathways, wound healing mechanisms, and the development of new drugs (Gupta et al., 2022).

Researchers developed a 3D computational model of a wound and the layers of skin to calculate the force needed for interrupted sutures. The force needed for each stitch in the presence of other sutures in the seven phases was estimated using a unique suture-pulling approach. Suture forces were calculated after a successful simulation of wound closure. The centre suture had a maximum force requirement of 3.7 N, which was roughly four times the force needed to place the first stitch (0.9 N). This will help in planning automated skin sutures in robotic operations, but it can also be applied in the future to sutures in other soft tissues or in routine excisions used to remove malignant tumors and lesions (Chanda et al., 2017).

The mechanical stimulation of the wound tissue alters the biochemical, mechanical, and cosmetic characteristics of resulting scar. Equi-biaxial stretch has been demonstrated to produce tissue with mechanical properties more nearly approaching the natural dermis, in contrast, to stretch across a wound during healing, which is known to emphasize alignment and stiffening of the tissue. For these researchers developed a technique to apply biaxial strain (equi-biaxial and strip biaxial) to model wound healing tissue in vitro to explore this phenomenon in a controlled experimental setting. A commercial vacuum-driven cell-stretching equipment was altered to apply in vitro cyclic biaxial strain to fibrin gels populated with fibroblasts to explore this phenomenon in vitro. Instead of multiactuator and inflation systems, this system was utilized as it provides a sterile environment for a long-term controlled dynamic culture of samples. Before manufacturing various platen, shapes and testing them in the actual system, finite element analysis was used to obtain theoretical strain distributions. This tool will make it possible to examine how biaxial stretch affects tissue models used to study wound healing (Holmes et al., 2005).

To attempt biomechanical modelling of damaged skin, researchers simulated several clotting stages to identify the impact of the skin's natural biaxial stress. They modelled realistic clinical wounds, particularly to patients using a FE framework to investigate the resulting stresses in the skin and the clot. The findings of this study will aid in evaluating any patient-specific wound healing effectiveness to allow preoperative planning, future applications in robotic surgeries, and the selection of wound closure methods that may result in the least amount of skin overstretching and rupture potential (Singh et al., 2022).

A combination of an experimental and a numerical technique may make for a useful method to characterize the mechanical behaviour of composite hydrogel films for use in wound healing applications. Karavasili et al. (2020) combined the nanoindentation measuring method with a finite element model (FEM) simulation to determine the material properties of composite hydrogels. The accuracy of the mechanical measurements was confirmed by a finite element model, illustrating the feasibility of combining experimental data with FEA simulations to more correctly establish the mechanical properties of compliant materials and further program their mechanical performance. These composite biomaterials have the potential to be effective in wound healing applications.

5.4 *Microneedle Insertion Analysis*

The simulation depicts the performance of the proposed device with respect to time. The development of the physical system is made easier by simulation, which is utilized in various contexts including the design, optimization, and testing of micro- and nanoscale devices. A simulation run in ANSYS can be used to analyse microneedle insertion with a model of the skin layers.

Researchers developed a skin insertion model for human skin and a microneedle in ANSYS AUTODYN. A skin layer model consisting of skin layers (stratum corneum, viable epidermis, and dermis) mimicking human skin characteristics (layer thickness and hypo-elastic characteristics) demonstrating the effect of microneedle insertion was analysed using the SOLID 186 design element. The mechanical parameters comprising stress, strain, and lateral deflection of each layer of skin were investigated. They concluded that a microneedle may puncture the three layers of skin under observation with a force of between 0.4 and 0.9 N (Tayyaba et al., 2020).

6 Conclusion

This chapter presents a basic introduction to the anatomy and physiology of integumentary systems to perform computer modelling. Heat transfer and fluid flow processes can be evaluated using CFD. Applications of CFD in health and biomedical research have recently advanced quickly. Because it is possible to reduce prototype construction and experimentation, CFD is both time and money efficient. Because there is still a lack of fundamental knowledge about the harmonious interactions between various cell types, intricate cell signalling networks, and mechanical feedback loops that change in space and time during healing, obtaining perfect skin regeneration after injury remains difficult. Hence, there is a need to design a computational model that combines insights from continuum mechanics, growth and remodelling, and systems biology regulatory networks to mimic the healing of wounds.

References

Casale, C., Imparato, G., Urciuolo, F., & Netti, P. A. (2016). Endogenous human skin equivalent promotes in vitro morphogenesis of follicle-like structures. *Biomaterials, 101*, 86–95.

Cavicchi, A., Gambarotta, L., & Massabò, R. (2009). Computational modelling of reconstructive surgery: The effects of the natural tension on skin wrinkling. *Finite Elements in Analysis and Design, 45*(8–9), 519–529.

Chanda, A., & Unnikrishnan, V. (2017). A realistic 3D computational model of the closure of skin wound with interrupted sutures. *Journal of Mechanics in Medicine and Biology, 17*(01), 1750025.

Flynn, C. (2010). Finite element models of wound closure. *Journal of Tissue Viability, 19*(4), 137–149.

Gupta, S., Patel, L., Mitra, K., & Bit, A. (2022). Fibroblast derived skin wound healing modelling on chip under the influence of micro-capillary shear stress. *Micromachines, 13*(2), 305.

Henrot, P., Laurent, P., Levionnois, E., Leleu, D., Pain, C., Truchetet, M. E., & Cario, M. (2020). A method for isolating and culturing skin cells: Application to endothelial cells, fibroblasts, keratinocytes, and melanocytes from punch biopsies in systemic sclerosis skin. *Frontiers in Immunology, 11*, 566607.

Holmes, M., Dufour, D., Kahan, M., Traynor, K., & Billiar, K. (2005, April). A method for applying strip biaxial stretch to cultured tissues. In *Proceedings of the IEEE 31st Annual Northeast Bioengineering Conference, 2005.* (pp. 168–169). IEEE. *Journal of Heat Transfer, 137*(12).

Karavasili, C., Tsongas, K., Andreadis, I. I., Andriotis, E. G., Papachristou, E. T., Papi, R. M., Tzetzis, D., & Fatouros, D. G. (2020). Physico-mechanical and finite element analysis evaluation of 3D printable alginate-methylcellulose inks for wound healing applications. *Carbohydrate polymers, 247*, 116666.

Makvandi, P., Shabani, M., Rabiee, N., Anjani, Q. K., Maleki, A., Zare, E. N., Sabri, A. H. B., De Pasquale, D., Koskinopoulou, M., Sharifi, E., & Sartorius, R.(2023). Engineering and development of a tissue model for the evaluation of microneedle penetration ability, drug diffusion, photothermal activity, and ultrasound imaging: a promising surrogate to ex vivo and in vivo tissues. *Advanced Materials*, 2210034.

McLafferty, E., Hendry, C., & Farley, A. (2012). The integumentary system: anatomy, physiology and function of skin. *Nursing Standard (Through 2013), 27*(3), 35.

Mercuri, M., & Fernandez Rivas, D. (2021). Challenges and opportunities for small volumes delivery into the skin. *Biomicrofluidics, 15*(1), 011301.

Mohizin, A., Roy, K. R., Lee, D., Lee, S. K., & Kim, J. K. (2018). Computational fluid dynamics of impinging microjet for a needle-free skin scar treatment system. *Computers in Biology and Medicine, 101*, 61–69.
Narasimhan, A., & Joseph, A. (2015). Porous medium modelling of combined effects of cell migration and anisotropicity of stratum corneum on transdermal drug delivery.
Piipponen, M., Li, D., & Landén, N. X. (2020). The immune functions of keratinocytes in skin wound healing. *International Journal of Molecular Sciences, 21*(22), 8790.
Ponmozhi, J., Dhinakaran, S., Varga-Medveczky, Z., Fónagy, K., Bors, L. A., Iván, K., & Erdő, F. (2021). Development of skin-on-a-chip platforms for different utilizations: Factors to be considered. *Micromachines, 12*(3), 294. Programmable and skin temperature–activated electromechanical synergistic dressing for effective wound healing. *Science Advances, 8*(4), eabl8379.
Rehfeld, A., Nylander, M., Karnov, K., Rehfeld, A., Nylander, M., & Karnov, K. (2017). The integumentary system. *Compendium of Histology: A Theoretical and Practical Guide*, 411–432.
Reid, L. (2021). An introduction to biomedical computational fluid dynamics. *Biomedical Visualisation:, 10*, 205–222.
Richardson, M. (2003). Understanding the structure and function of the skin. *Nursing times, 99*(31), 46–48.
Rodríguez, M. R., Otero, A. T., Acha, L. Y., Gutiérrez-Rivera, A., Paredes, J., Izeta, A., & Aldazabal, J. (2017). Study and analysis of the effects of psychological stress, mechanical stresses and wound shape in wound healing process both in vivo and in silico. *Statistics, 100*, 1.
Romanovsky, A. A. (2014). Skin temperature: Its role in thermoregulation. *Acta Physiologica, 210*(3), 498–507.
Ruela, A. L. M., Perissinato, A. G., Lino, M. E. D. S., Mudrik, P. S., & Pereira, G. R. (2016). Evaluation of skin absorption of drugs from topical and transdermal formulations. *Brazilian Journal of Pharmaceutical Sciences, 52*, 527–544.
Singh, G., & Chanda, A. (2022). Biomechanical modelling of progressive wound healing: A computational study. *Biomedical Engineering Advances, 4*, 100055.
Tayyaba, S., Ashraf, M. W., Tariq, M. I., Nazir, M., Afzulpurkar, N., Balas, M. M., & Mihalache, S. F. (2020). Skin insertion analysis of microneedle using ANSYS and fuzzy logic. *Journal of Intelligent & Fuzzy Systems, 38*(5), 5885–5895.
Tepole, A. B. (2017). Computational systems mechanobiology of wound healing.
Yang, G., Long, H., Wu, J., & Huang, H. (2008, May). A novel electrical field bioreactor for wound healing study. In *2008 International Conference on BioMedical Engineering and Informatics* (Vol. 2, pp. 548–552). IEEE.
Yao, G., Mo, X., Yin, C., Lou, W., Wang, Q., Huang, S., Mao, L., et al.: A programmable and skin temperature–activated electromechanical synergistic dressing for effective wound healing. *Science Advances 8*(4), eabl8379 (2022)
Yousef, H., Alhajj, M., & Sharma, S. (2022). *Anatomy, skin (integument), epidermis*. StatPearls. Treasure Island.

Chapter 6
Role of Computational Fluid Dynamics in Cancer

1 Introduction

Millions of people die from cancer every year, and despite considerable medical advancements, there are still a lot of challenges that need to be taken into account to improve cancer treatment. Oncological research is concentrating on finding novel and effective therapies that can mitigate significant side effects brought on by traditional treatments. The main objective of this chapter was to comprehend how computational fluid dynamics (CFD) can be used to assess the accuracy and efficacy of cancer treatment strategies. The study helps in understanding the effectiveness of several cancer treatment methodologies. Additionally, it offers a platform for several experiments that would be extremely challenging to carry out in a real-world setting. This chapter focuses on CFD-based simulation models on cancer treatment approaches. Using CFD has greatly decreased the time-consuming expenditures of doing physical experiments. Many researchers offer a clinical viewpoint on CFD modelling and analyse cancer-related CFD models from a fluid mechanics approach. It goes through how well CFD can currently simulate cancer and potential future directions in this area. Overall, it can be said that CFD can serve as a physical examination substitute because it produces accurate and affordable results. However, validation with in vivo and in vitro tests is required to prove the accuracy of CFD simulations as a research tool.

2 Background

The fundamental characteristics of cancer, a common disease that is actively spreading across the globe due to altered dietary and living habits, are unchecked cell proliferation and aberrant cell dissemination. Normal body cells typically develop,

S. E. Jujjavarapu et al., *Computational Fluid Dynamics Applications in Bio and Biomedical Processes*, https://doi.org/10.1007/978-981-99-7129-9_6

divide, and eventually die in a controlled manner, but the distinguishing characteristic of these cancer cells is that they continue to proliferate and create aberrant cells, which results in the development of a mass of tissue known as a tumor. Tumor can be categorized into two categories, namely benign and malignant. A benign tumor poses fewer risks to a patient's life because it does not spread to other body regions. The lymphatic system and bloodstream allow malignant tumor to easily invade adjacent tissues and organs before disseminating to other areas of the body. Metastasis is the term used to describe the spread of tumor through the blood or lymphatic system to other parts of the body, which can result in serious health issues or even death. There are different types of cancer, each having its characteristics and risk factors and the treatment for each can be multiple or individual techniques depending on the type and stage of the cancer.

CFD, a subfield of fluid mechanics, uses numerical techniques and algorithms to simulate and analyse the behaviour of fluid flows (Fig. 1). The governing equations for fluid flow are solved via computer simulations. This branch of study helps to study and predict the complex behaviour of fluids in different environmental conditions. CFD is a powerful tool for providing valuable insights and is a suitable source in conjunction with other research methods and clinical data to enhance the understanding of cancer and help in the improvement of its treatment strategies. CFD in cancer research and treatment has particularly aided in the field of medical imaging and therapy planning. For cancer analysis, CFD aids in tumor blood flow analysis, drug delivery optimization, radiation therapy planning, thermal ablation guidance, personalized medicine, and in silico testing. There are three levels at which heat therapy has an impact: cellular and molecular, tissue, and systemic. These findings provide special benefits when thermal therapy is combined with chemotherapy, radiation therapy, immunotherapy, and other minimally invasive therapies.

Over the last few years, CFD has actively contributed to the understanding and treatment of cancer across the research community. Through various studies performed by scientists across the globe, it is found that studies like tumor vasculature and drug delivery have investigated the blood flow patterns, interstitial fluid flow, and drug transport within tumor to optimize the strategies for drug delivery, microfluidics, and cancer cell behaviour to study the behaviour of cancer cells in different microenvironments that mimic the aspects of the tumor environment. Tumor growth and angiogenesis have been used to study the tumor growth and blood vessel development within these tumor, and biomechanics of cancer cells which helps to study the biomechanical behaviour of cancer cells like the cell deformation, migration, and interaction with the extracellular matrix.

CFD has been a useful tool in understanding cancer cell mechanics in the past. It has been a promising tool to be used in future as well to guide its development and understanding. This chapter will provide insight into possible and useful applications of CFD in cancer research that scientists from around the world have thrown light upon so far.

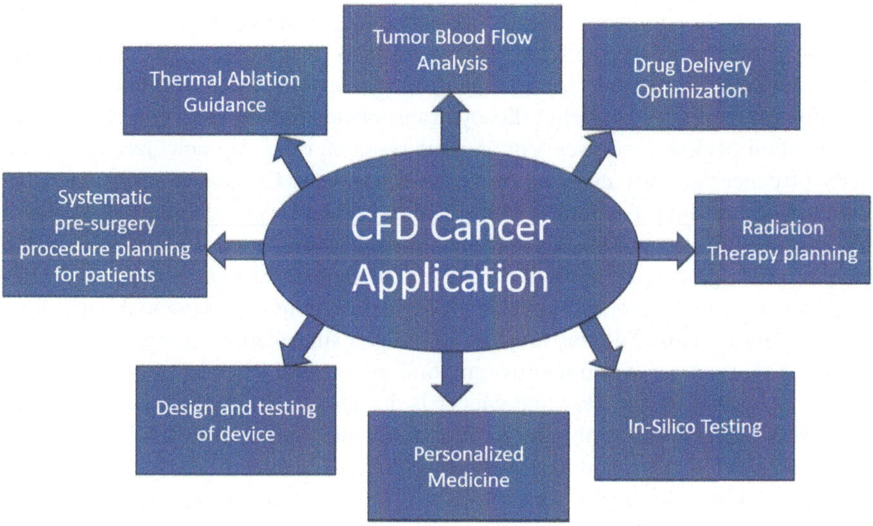

Fig. 1 Schematic diagram of cancer research involving the use of CFD

3 Fluid Mechanics Study of Cancer

The fluid mechanism of cancer potentially refers to the application of fluid mechanics principles and techniques to study the behaviour and dynamics of cancerous tumor within the domain of fluid flow. The examination of these fluid mechanism patterns of growth, progression, and treatment of tumor in the microenvironment helps in a detailed analysis of cancer. Tumor angiogenesis, tumor microenvironment, metastasis, drug delivery, treatment planning, and biomechanical modelling are common studying concepts that are being looked at present in the research world concerning cancer as this subject helps in integrating the concepts of fluid dynamics, oncology, bioengineering, and computational modelling.

Fluid mechanics has a significant impact on cancer biology, affecting the development, spread, and treatment of cancer. The tumor microenvironment experiences higher interstitial pressure due to the tumor's erratic and leaky vasculature. Transporting oxygen, nutrients, and medications to tumor is largely dependent on blood arteries. Tumor growth, metastasis, and medication distribution are all governed by mechanisms of multiscale flow-structure interaction. Utilizing computational and experimental methods, these flow-mediated pathways can be investigated and modelled to aid in the detection, prognosis, and treatment planning of cancer (Koumoutsakos et al., 2013). Understanding fluid mechanics in cancer can help develop creative methods for fighting the disease. The significance of fluid mechanics in the onset, progression, and metastasis of cancer is thoroughly explored in this paper, along with experimental diagnostics, computer modelling, and therapeutic strategies. Several flow-mediated processes in cancer are examined, including intravasation, aberrant molecular diffusion during the development of tumor, endothelial cell migration, nutrition transport during angiogenesis, vascular remodelling brought on by flow-induced stressors, and tumor cell metastasis (Rajput et al., 2023; Steuperaert et al., 2019). In therapeutic tactics like drug delivery to tumor and vascular regularization to increase treatment efficacy, transport phenomena are also crucial. The combination of flow processes with genetic research opens up new avenues for the study of cancer and the creation of potent treatments. Collaboration between the communities of fluid mechanics, biology, nanotechnology, and cancer medicine is necessary for this integration. Even though there has been a lot of progress in understanding the significance of fluid mechanics in cancer, there is still a lot of room for improvement. To properly measure these interactions, experimental, analytical, and simulation methodologies must be developed since cancer comprises complex processes that cut across numerous temporal and spatial domains. Future research will be motivated by the idea that cancer is a systemic disease with a strong fluid mechanics component, combining experimental and computational fluid mechanics with genetic, molecular, and nanotechnology advancements.

Understanding different facets of cancer biology requires knowledge of fluid dynamics or fluid mechanics:

(a) Hemodynamics: It is the study of blood flow within the body including the circulation through the blood vessels. In cases of cancer, it is important to understand the hemodynamics of tumor blood vessels as the changes in the blood flow patterns such as increased blood vessel tortuosity, irregular branching, and heterogeneous perfusion which are the characteristic transformation of a tumor vasculature. Computational modelling can aid in the investigation of these complex flow patterns and the forces by simulations and mimicking the microenvironment pattern for cancer studies.
(b) Interstitial fluid flow: It plays a very important role in tumor progression, invasion, and metastasis. Modelling of such structures helps in understanding the nutrient transport, waste products, and signalling of molecules within the tumor as well as the interplay between the interstitial fluid flow and the cell behaviour.
(c) Extracellular matrix (ECM) Mechanics: The structural properties of the extracellular mechanics influence tumor progression and invasion. The fluid mechanics principles help in studying rheology and viscoelasticity, and this helps in characterizing the mechanical properties of the ECM. Modelling in turn aids in providing the understanding and insight into the stiffness, porosity, and composition of the ECM which later affects cell migration, tumor growth, and response to therapy.
(d) Lymphatic Fluid Flow: It is related to the lymphatic system and is responsible for draining fluid and waste products from the tissues. The lymphatic system can play a significant role in tumor metastasis, thus the understanding of the lymphatic fluid and its interaction with the cancer cells is important in studying the spread of cancer to lymph nodes or any other distant site.
(e) Cerebrospinal fluid flow (CSF): This fluid flow can be altered in case of brain tumor. Different fluid dynamics studies have been conducted to investigate the

impact of tumor growth on CSF flow in tumor progression and therapeutic interventions.

3.1 Hemodynamics Transport Study of Cancer

The hemodynamics transport study of cancer includes the analysis of fluid flow within the vasculature of tumor and its implications for cancer progression and treatment. This subject of interest helps in understanding the transport of nutrients, oxygen, waste products, therapeutic agents, and other solutes within the tumor microenvironment. This study of the hemodynamics of cancer provides insight into the complex interplay between fluid dynamics and tumor physiology. The tumor blood flow is the analysis of the blood flow characteristics within the tumor vasculature. It involves the investigation parameters including blood velocity, pressure, and flow distribution. The blood flow patterns in the tumor help in identifying the regions of hypoxia and assessing the efficiency of nutrient delivery and waste removal. Angiogenesis and vascularization contribute to the understanding of tumor angiogenesis which simulated the growth of new blood vessels. This study helps in identifying potential targets for antiangiogenic therapies. Drug delivery optimization is also studied through the hemodynamic mechanisms as it assesses the distribution and penetration of therapeutic agents within the tumor microenvironment. It also assists in understanding the blood flow effect on drug transport which in turn helps to optimize drug administration strategies and improve the efficacies of cancer therapy. Microvascular permeability refers to the ease with which solutes pass through the blood vessel walls. Alterations in this property of tumor influence the extravasation of cells and the delivery of therapeutic agents. The factor analysis for the factors enhances drug delivery and inhibits metastasis. Fluid cell interactions mean the interactions between the flowing blood and tumor cells, and this includes the examination of cell adhesion to the vessel walls, mechanical deformation of cells to the response of fluid shear stress, and the impact of flow-induced forces on cell signalling and behaviour. These affect the tumor cell dissemination, invasion, and response to therapy. CFD is used to model and analyse tumor hemodynamics. This tool provides a quantitative understanding of the blood flow patterns, fluid pressures, and solute transport within the tumor vasculature.

Integrating the data obtained from medical imaging and mathematical models helps the researchers to simulate and predict the hemodynamic behaviour in patient-specific tumor scenarios. The modelling in CFD can help to study the following:

(a) Blood Flow Simulations: CFD allows the simulation and analysis of complex blood flow patterns within the tumor vasculature. Solving the governing equations of fluid dynamics provides information on blood velocity, pressure distribution, and flow direction. This analysis helps in understanding the hemodynamics of tumor vessels and identifying regions of high or low flow, areas of stagnation, and regions that are prone to hypoxia.

(b) Hemodynamic Parameters: Computational modelling enables the calculation of various hemodynamic parameters which are important for understanding tumor physiology. The parameters include wall shear stress and pressure gradients.
(c) Treatment Planning and Assessment: CFD simulations help in designing the prototype of treatment and assessment by considering the hemodynamic factors included in the therapy. For instance, CFD modelling for radiation therapy can account for the movement of normal and tumor tissues due to respiratory motion, helping in optimizing the treatment plans and minimizing healthy tissue damage. CFD can also help in evaluating the effectiveness of treatment modalities such as radiofrequency ablation or target drug delivery systems by the prediction of fluid flow and solute transport.
(d) Patient-Specific Modelling: This describes CFD modelling that combines the developed computer models with patient medical information, such as computed tomography (CT) or magnetic resonance imaging (MRI) pictures. This provides the opportunity for enabling the personalized simulations of tumor hemodynamics, considering individual vascular geometries and physiological characteristics. These models provide insights into tumor-specific hemodynamics and guide personalized treatment decisions.

Thus, computational modelling of the hemodynamics transport study of cancer provides a detailed understanding of the fluid flow patterns, quantifies hemodynamic parameters, optimizes the drug delivery strategies, and supports treatment planning for improved patient outcomes.

A patient-specific CFD model for blood flow and interstitial transport in breast cancer is being developed as part of this project using quantitative magnetic resonance imaging (MRI) data (Wu et al., 2020). Using magnetic resonance imaging (MRI) data relevant to the patient, a computational model was developed to mimic the blood supply and interstitial fluid environment in breast cancers. To diagnose a patient, provide medication, and evaluate a patient's response to treatment, it is crucial to understand how flow via tumor-associated arteries interacts with the interstitium. The pressure and flow fields inside the breast are mapped using this method, which is a ground-breaking attempt to merge CFD and patient-specific MRI data. This approach offers crucial insights for the research of breast cancer. With the aid of diffusion-weighted magnetic resonance imaging (DW-MRI) and dynamic contrast-enhanced magnetic resonance imaging (DCE-MRI) data, tumor-related vascular-interstitium geometry and material properties were developed. This information was then used to restrict the CFD simulations to identify the interstitial transport properties and the tumor-associated blood supply. A statistical analysis reveals significant variations in tumor-associated interstitial flow velocity, blood pressure, and vascular extraction rate between malignant and benign lesions. Our image-based model approach offers a non-invasive method of evaluating flow and pressure fields in breast tumor by having the capacity to quantitatively define breast cancer. This novel approach describes the combination of patient-specific MRI data and CFD analysis to define the hemodynamics of breast cancer. By combining additional patient-specific information and considering arbitrary tumor geometries, diagnostic accuracy can be increased. The findings show that there are major hemodynamic differences between benign and malignant lesion groups. When compared to benign lesions, malignant lesions show higher interstitial flow velocity, increased blood pressure heterogeneity, and higher rates of vascular extraction. These results demonstrate the promise of contrast agent pharmacokinetics as a cutting-edge method for assessing breast cancer.

3.2 Interstitial Fluid Transport Study of Cancer

Interstitial fluid transport study of cancer refers to the study of the transport and dynamics of fluid within the tumor microenvironment, specifically the fluid that exists between the cells and blood vessels. This interstitial fluid plays a vital role in different physiological phenomena including nutrient and waste exchange, cell signalling, and the spread of cancer cells. The study of interstitial fluid covers various studies in cancer research including interstitial fluid pressure (enhanced pressure is a common characteristic of solid tumor, and it is associated with poor drug penetration, reduced oxygen delivery, and impaired immune response), interstitial fluid flow, extracellular matrix (it affects the fluid flow and determines the spatial organization of cells within the tumor, the interaction between ISF and ECM helps to understand the effects of tumor microenvironment on cancer cell behaviour), convection and diffusion (contributes to the transport of molecules and drugs within the tumor), heterogeneity and transport barriers, and mathematical modelling. Computational modelling for the study of interstitial fluid transport in cancer will help to provide a quantitative and detailed analysis of fluid flow and solute transport within the tumor microenvironment.

The tumor microenvironment can be simulated using CFD, and by resolving the equations regulating the fluid dynamics, CFD models can forecast fluid velocities, pressures, and flow patterns. Identifying areas of fluid stasis or high flow and determining the influence of tumor features on fluid transport are made easier with an understanding of the dynamics of the ISF flow. Additionally, this application offers simulations that make it possible to compute the ISF pressure and flow distribution within the tumor. The transport of oxygen and nutrients to the tumor cells as well as the penetration of drugs are impacted by elevated interstitial fluid pressure. The interstitial fluid pressure and the associated factors can be measured using the CFD models.

The study of the transport of solutes and therapeutic agents can be done using CFD models as they can simulate the transport of solutes such as oxygen, nutrients, waste products, and therapeutic agents within the tumor interstitium. Incorporation of the convection and diffusion equations in CFD can predict the transport and distribution of solutes based on the fluid flow patterns and the properties of the tumor microenvironment. The evaluation of drug delivery strategies, optimized treatment protocols, and the understanding of the spatial heterogeneity of solute transport can be used to evaluate the treatment strategy. CFD simulations assist in studying the impact of tissue permeability and transport barriers on interstitial fluid transport by incorporating information on the ECM and cellular structures. The CFD model can assess the hindrance to fluid and solute movement caused by compressed blood vessels, high ECM density, and cellular structures; the obtained information aids in the understanding of the limitations and heterogeneity of solute transport within the tumor.

Computational modelling for studying interstitial fluid transport in cancer can allow for patient-specific modelling by integrating imaging data with computational models. This allows the designing of the prototype for personalized treatments by considering individual tumor geometries, physiological properties, and treatment plans. The detailed information obtained by the computational study of the various interstitial fluid transport in cancer helps in optimization of the further treatment procedures.

Researchers also worked on a mathematical model for interstitial fluid flow in physiological systems with solid tumor (Soltani et al., 2011). The model is applied to a predetermined tumor geometry and makes use of the laws of conservation of mass and momentum. The interstitial fluid pressure and velocity are determined numerically. Simulations show that the interstitial pressure distribution causes the drug particle distribution in uniformly perfused tumor without necrotic patches to be non-uniform. The study demonstrates that smaller tumor below the essential size have simpler drug transport to the centre than larger tumor by identifying crucial tumor and necrotic radii. Additionally, increased medication concentrations and therapeutic benefits are accomplished within particular ranges of these critical diameters due to lower interstitial fluid pressure. This work focuses on the interstitial fluid flow-based numerical analysis of medication distribution in solid tumor. According to the findings, uniformly perfused tumor' uneven medication distribution is primarily caused by high interstitial pressure. The distribution of the interstitial fluid's pressure and velocity is determined using numerical solutions to the governing equations and the presumption that drug particles flow through it. The tumor centre has the maximum interstitial pressure, which decreases towards the periphery, according to a comparison with experimental data. The study describes two additional parameters that affect how quickly drugs can reach the tumor centre: the critical tumor radius and the critical necrotic radius. Drug transmission is simpler in smaller tumor compared to larger tumor because they have lower interstitial fluid pressure. The study also demonstrates that the maximal pressure inside the tumor is reduced when a necrotic zone is present. To accommodate anisotropic tissues and two-phase flow modelling, which is important for medications with nanoparticle sizes comparable to capillary diameters, the numerical model can be expanded. The results show the importance of interstitial pressure in drug absorption into solid tumor as a whole and offer guidance for future modelling of complex tissue scenarios.

4 Diagnosis of Cancer Metastasis

Metastasis, a crucial step in the progression of cancer refers to the detachment of the primary tumor, entering the bloodstream or the lymphatic system and travelling to distant organs or tissues once they reach the new location, these cells can invade or grow forming secondary tumor which are similar to the original cancer. Because not all cancer cells have the ability to spread to other parts of the body via lymphatic or blood vessels, and because some of these cells enter the bloodstream or lymphatic system and are either killed by the body's immune system or are unable to adapt to the new environment to survive, the removal of primary tumor from their original sites is a difficult and multistep procedure that does not guarantee the development of secondary tumor. Because it can cause cancer to spread throughout the body and become harder to treat, this trait is a crucial characteristic of malignant or cancerous and is to blame for the majority of cancer-related deaths. Typically, chemotherapy, targeted therapy, immunotherapy, surgery, radiation therapy, and palliative care are used to treat metastatic cancer. Imaging procedures like X-rays, CT scans, MRIs, PET scans, bone scans, and ultrasounds are frequently used to identify cancer. Biopsy, blood tests, endoscopy, and molecular profiling are also used often by healthcare professionals for cancer diagnosis and to determine the most appropriate diagnostic method for each case.

CFD is not a direct diagnostic tool for cancer; however, this tool can be used to model and understand the fluid dynamics such that it can aid the process of diagnosis of cancer metastasis. Drug delivery optimization, blood flow analysis, and radiation therapy planning are some of the phenomena which can be studied using CFD such that it helps to research further and provide a treatment plan for the diagnostic tool of cancer metastasis as this diagnosis relies on comprehensive assessment using various clinical and imaging techniques.

5 Pharmacokinetics Study

Pharmacokinetics, a branch of pharmacology, focuses on the study of how the body is affected after drug administration. This entail looking into how the body absorbs, distributes, metabolizes, and eliminates drugs. These processes influence the concentration of the drug in different tissues and organs over time. This study involves the collection of data on drug concentrations in biological samples at different points in time after the drug administration. This information is used to characterize the appropriate dosage regimen, understand drug interactions, evaluation of drug safety, and optimize therapeutic outcomes.

An overview of the fundamentals and constraints governing medication delivery to solid tumor may help in improving treatment efficiency. As can be seen, the vasculature of normal and tumor tissues differs, and some factors affect how medications traverse tumor vasculature and enter neighbouring tumor tissues. Numerous factors can affect the absorption, retention, and accumulation of drugs inside tumor. For cancer therapy to be effective, therapeutic medications must efficiently reach tumor cells. This highlights the requirement for a deeper comprehension of these mechanisms to improve medication delivery in solid tumor. The delivery of macromolecules or the passage of drugs through solid tumor can be impacted by several factors. The tumor's diverse physiological characteristics as well as its physicochemical characteristics, including its binding to extracellular and intracellular components and diffusivity, all come into play. Angiogenesis, vascular distribution, tumor cell density, lymph flow, microvessel density, interstitial fluid pressure (IFP), and stromal tissues are a few examples of the elements that affect drug transport. Drug transport in tumor is a dynamic process since these biological characteristics are subject to alter over time and as a result of drug-induced side effects. Compared to small molecule medications carried by simple diffusion, high-binding pharmaceuticals rely on convection, which is highly impacted by dynamic biologic features. For instance, until apoptosis takes place, the transport of an apoptosis-inducing medication may be slower in a tumor with a high cell density. It's significant to note that various biologic variables associated with tumor have conflicting and interrelated impacts on medicine delivery and transport. In addition to improving drug delivery via perfusion, higher blood pressure also causes an increase in IFP, which inhibits interstitial space transport. Therefore, the many different factors involved should be taken into account in comprehensive studies on drug delivery and transport. While diffusion coefficients and IFP have received the majority of attention, spatial drug distribution within tumor has received less consideration. Effective cancer treatment depends on achieving uniform medication distribution throughout the vascular and avascular sections of a tumor. With better imaging capabilities to see drug distribution in vitro and in vivo, future research should focus on comprehending and removing obstacles to drug transport within tumor. In solid tumor, a variety of physicochemical and biologic variables, some of which are dynamic and alter with time and therapy, affects medication delivery, transport, and spatial distribution. With the introduction of novel treatment drugs, an understanding of these characteristics can help in the development of therapeutic methods for both passive and active tumor targeting.

Similar to cancer metastasis diagnosis, CFD cannot be employed to study the pharmacokinetic study directly but through other methods it can be used to contribute to the pharmacokinetic contributions (Fig. 2). CFD simulations for drug delivery systems (such as inhalers, nebulizers, patches, or implants) can be made by incorporating the physiological models and computational fluid dynamics; this helps the researchers in evaluating the behaviour of the drug within the delivery device and predict drug deposition patterns in the patient body. The obtained information helps in optimizing the drug formulations and delivery strategies to enhance drug absorption and achieve the desired pharmacokinetic profiles. This quantitative method can also be employed in analysing the drug deposition and distribution within the lungs following inhalation. The modelling of respiratory system particle movement, deposition, and airflow patterns can help predict the regional distribution of inhaled drugs and assess factors influencing drug absorption, such as particle size, inhalation technique, or airway geometry. The information gained for pulmonary drug delivery

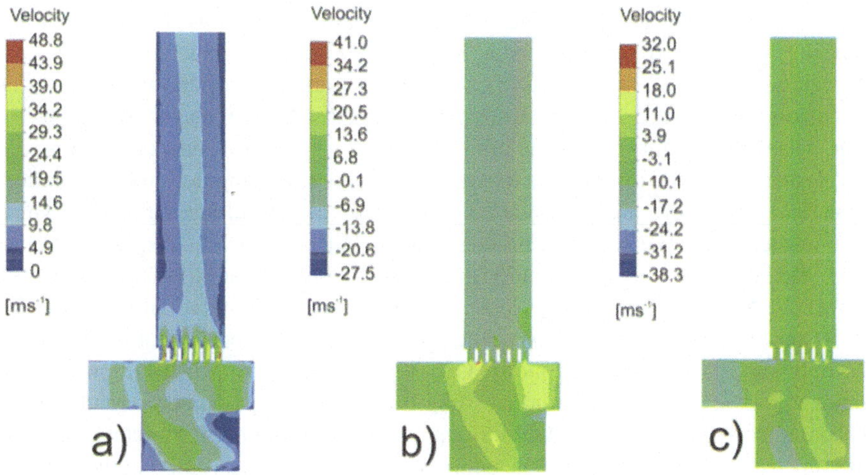

Fig. 2 Figure showing **a** velocity magnitudes, **b** tangential, and **c** radial velocities of jet-milled samples for the xz plane calculated by CFD simulation (Vulović et al., 2018)

computational modelling contributes to the understanding of drug pharmacokinetics in the lungs and assists in its development of it. CFD can also be employed to understand the drug–drug interactions within the body. The simulation of the distribution and transport of multiple drugs in the bloodstream can be studied using CFD to understand how the drugs interact with each other and potentially influence their pharmacokinetic behaviour. This information predicts the impact of drug combinations on drug concentrations, metabolism, and elimination. These CFD simulations also provide insights into the tissue perfusion which affects the drug distribution and exposure in different organs and tissues. This modelling of blood flows through the tumor vasculature and incorporating drug properties is useful in understanding the pharmacokinetic variations among different organs and assists in predicting the drug concentrations at target sites.

Thus, we can see that even though CFD cannot be a primary tool for the evaluation of the pharmacokinetic behaviour of drugs. This study typically involves experimental data, pharmacokinetic modelling, and clinical observation to evaluate drug absorption, distribution, metabolism, and elimination. By offering thorough details on the transport and distribution of pharmaceuticals within physiological systems or drug delivery devices, CFD complements existing approaches.

Understanding the interstitial fluid pressure and the heterogeneity of the tumor's vasculature, which are the main inhibitors of drug penetration and cell killing, is made possible by the study the researchers conducted to comprehend the CFD analysis of drug uptake and their elimination through the vascularized tissue (Shojaee et al., 2019). To determine the steady-state interstitial fluid pressure and velocity, they used Starling's and Darcy's laws. In this experimental design, alternative post-drug administration dosages and time windows are also examined. Assuming that capillary walls are leaky, Starling's law is applied to determine the trans-vascular fluid velocity. Both healthy tissues and tumor contain qualities that are regarded as being indicative of porous media. According to Darcy's law, the velocity and momentum equation can be used to calculate the pressure drop in the domain. The two distinct explanations for the drug transport pathways are convection and diffusion. The diffusion term faces the tumor from the inside out. Convection, however, travels away from the tumor in an outward direction. The computer model's applications for studying this are then focused on the interstitial fluid pressure (IFP), which is connected to the intravascular blood pressure (IBP) by Starling's equation. In this work, the governing Darcy's equation and steady-state laminar blood flow were both COMSOL Multiphysics-solved problems. The quantity of interstitial fluid velocity (IFV) and IFP is implemented using the transient drug transport equation to resolve the bolus and continuous equations. Overall, the medication distribution in solid tumor microvasculature was studied using this model. The boundary conditions at the pressure inlet and output were thought to be the main determinants of blood pressure as a variable quantity. The entire model is based on simulating the delivery of doxorubicin to malignant tissue while incorporating randomly distributed vasculature that is denser towards the tumor. The outcomes demonstrated that the IFP rises as a result of the IBP. In contrast, in the case of bolus injection, it reached the pinnacle in 0.2 h with a rapid decline, leading to the conclusion that continuous is preferable to bolus mode. The interstitium's diffusion term, however, dwarfs the convection rate in importance.

Also, there is an investigation into drug delivery in a vascularized medium using these values in a transient drug transport equation including taking plasma concentration into account (Shojaee et al., 2019). According to the findings, cell death and medication penetration are hampered by high interstitial fluid pressure and vascular heterogeneity. The importance of pressure gradients and low convection rates in the tumor tissue was highlighted when various medication doses, intervals, and injection techniques were investigated. Three components make up the computational model employed in this study: a capillary network, tumor tissue, and normal tissue. In the geometry, normal and tumor tissue are represented by rectangular and semicircular shapes, respectively, with various features in the tumor capillaries. The model uses Darcy's equation and Starling's equation to determine the pressure and velocity of interstitial fluid. Three distinct grid arrangements are compared, and the one with 53,684 quadrilateral boundary layer elements and 557,773 triangular grids is chosen for investigation. Using a computational model with randomized vasculature that is thicker in the tumor area, it concentrates on simulating the transport of doxorubicin to malignant tissue. The results show that continuous infusion causes doxorubicin to remain in the tumor tissue for a longer period than bolus infusion. According to the study, a continuous strategy with an ideal dose and time is preferred for more effective tumor cell eradication.

Biophysical modelling is a powerful tool for pharmaceutical research and development because it offers a clear and thorough understanding of drug pharmacokinetics and therapeutic design. These models are capable of describing drug transport, distribution, and therapeutic effects by fusing physiological observations, pharmacological characteristics, and experimental data. This method is demonstrated in a case study employing an explicit geometry-based computed fluid dynamics model of the spinal canal from magnetic resonance imaging (Kuttler et al., 2010). The model, which was put into practice using reputable multiphysics software, offers insights into the variables affecting drug distribution following lumbar delivery. By linking drug delivery methods to the complex causal chain of drug distribution, this biophysical modelling strategy advances pharmaceutical research. Real geometry derived from MRI data was used to create a 3D computational fluid dynamics model of the spinal canal, which was then governed by transient Navier–Stokes equations. This model takes into account crucial information on the intrathecal distribution of medications following lumbar delivery. This model provides a thorough understanding of the transport mechanisms throughout the system through the integration of biophysical fluid flow dynamics principles and physiological traits of the cerebrospinal system. The simulation's findings have brought to light several crucial elements that affect how drugs are distributed inside the CSF. For instance, key factors in establishing the pattern of drug distribution include the speed and method of drug delivery, needle orientation, and volume of fluid administered. Laminar or turbulent flows may influence axial dispersion, but the basic driving factors of pulsation and breathing ultimately control the transport process. Drug administration in spinal cord injury and clinical trials may be affected by this biophysical modelling strategy practically. The model supports decision-making processes by modelling various administration scenarios, which eliminates the need for potentially unethical human dose experimentation. Furthermore, given the difficulties in generating consistent medication concentrations throughout the CSF, it gives clinicians confidence when deciding on the best administration mode for patients. Additionally, the drawbacks of traditional pharmacokinetic methods for spinal drug administration are explored, especially when drug exposure occurs elsewhere than at the injection site. The model provides a viable option for understanding medication distribution and concentration levels in the CSF over time at various spinal levels while taking into consideration physiological factors and biophysical principles (Fig. 3). Its usefulness for predicting individual dosing utilizing imaging and computational techniques is further increased by the integration of experimental data and enhancement of the model's functionality, including molecule binding and metabolism. The spinal CSF dynamics biophysical model offers a cutting-edge framework for comprehending medication distribution in the intrathecal region. It tackles issues with the dose-concentration-effect relationship and provides knowledge of the variables affecting medication absorption into the CSF, advancing model-based drug development and improving spinal drug delivery.

The transport of drugs and their effects during chemotherapy can be predicted using multiscale mathematical modelling (Kashkooli et al., 2020). The method accounts for drug binding, cellular uptake, fluid movement, drug transport, the shape of the capillary network in the extracellular matrix, and the necrotic core of the tumor. The results show a more accurate calculation of the percentage of cancer cells that were destroyed when compared to earlier models. The study examines how drug delivery to solid tumor is impacted by characteristics linked to tumor, such as necrotic core size and tumor size, dosage, binding affinity, and drug degradation.

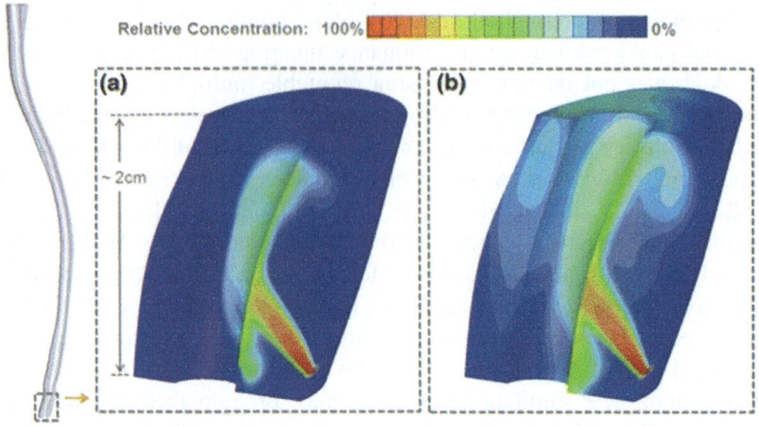

Fig. 3 Biosimulation of drug distribution using CSF flow was used to simulate an intrathecal injection at the lumbar location L2 (flow rate: 0.3 ml/s) at **a** 1 s and **b** 3 s (Kuttler et al., 2010)

The findings highlight the importance of tumor size and pharmacological properties in treatment efficacy and urge for personalized therapy based on these factors. The study offers knowledge of drug delivery systems and how they affect patient outcomes. The steps of drug delivery to solid tumor include drug injection, capillary transport, extravasation into tissues, interstitial transport, binding to cell receptors, and cellular absorption. The effect of various parameters on medication distribution in tumor and healthy tissues is examined using mathematical modelling. The research examines the spatiotemporal drug distribution using microvessel geometry taken from an image. This research focuses on the systemic administration of chemotherapeutic medications to solid tumor to minimize side effects on healthy organs and enhance therapeutic efficiency. A more accurate model of solid tumor is used to conduct an extensive and realistic investigation of interstitial and intravascular fluid flows (Fig. 4). The capillary network obtained from tumor imaging, the necrotic zone inside the tumor form, and the consequences of medication degradation are only a few examples of the variables the model takes into account. The size of the tumor, the necrotic region, the drug dosage, and the binding affinity to cell-surface receptors are a few of the variables that are examined in the study. Drug concentration is often found to be extremely low in necrotic areas, low in healthy tissue, and high in tumor. It has been found that tumor with fewer necrotic zones respond better to treatment. A 12-mm tumor is suggested as the ideal candidate for treatment since it is more effective in the tumor location and has less detrimental effects on healthy tissue than other tumor sizes. The study also emphasizes the significance of choosing a suitable therapeutic dosage based on the condition of the patient and the characteristics of the tumor. The effectiveness of the treatment in the tumor region initially increases with higher drug binding affinities to cell-surface receptors, but after a certain point, the effectiveness declines. Drug deterioration barely affects how well a treatment works. The findings demonstrate the need of balancing treatment effectiveness and adverse

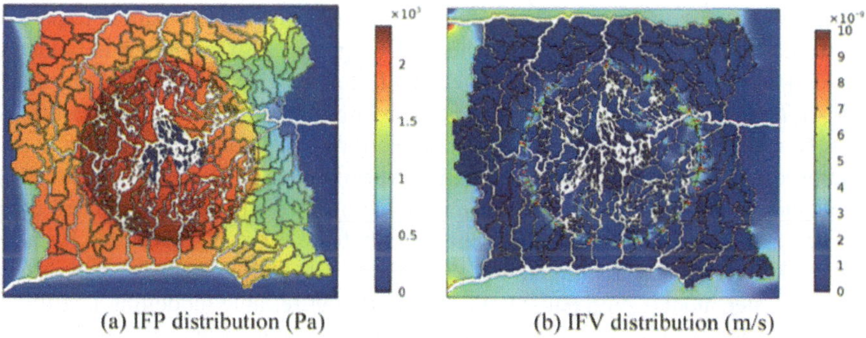

Fig. 4 Distribution of interstitial fluid pressure and interstitial fluid velocity in ECM of tumor with heterogeneous microvascular networks (Kashkooli et al., 2020)

effects by taking into account the most effective set of working parameters. Overall, the research aids in the development of more suitable medications and personalized treatment plans. The proposed computational approach applies to numerous drug classes and can help in therapy planning and treatment response assessment, taking into account the particulars of each patient.

Intricate physiological, metabolic, and biophysical processes that take place on multiple time and length scales are involved in the drug delivery to solid tumor. Controlled studies of the effects of various parameters on medication transport and efficacy are made possible by mathematical and computational modelling, which are unfeasible through testing alone (Zhan et al., 2018). Numerous mathematical models, such as compartmental models, transport models at various scales, and molecular dynamics models, have been created. These models support the optimization of medication delivery systems and the identification of limiting factors. They have been used in cutting-edge delivery systems like convection-enhanced delivery and delivery mediated by nanoparticles. In general, computer models help in the development of effective drug delivery methods and offer useful insights into drug transport in solid tumor. Drug delivery systems (DDS), such as catheter-needle systems for convection-enhanced distribution, polymeric implants, and carriers for nanoparticles, are developed and designed in large part by mathematical modelling. Drug loading capacity, bloodstream nanoparticle stability, and drug release profiles are only a few examples of the variables that affect how successful nanoparticle-based DDS is. Particularly helpful in comprehending these systems are molecular dynamics (MD) simulations, which offer thorough details on carrier structure, drug interactions, and nanoparticle production optimization. By providing a microscopic comprehension of findings and pointing out constraints, MD studies complement experiments. There are three types of approaches: atomistic, coarse-grained (CG), and dissipative particle dynamics (DPD), with hybrid atomistic-CG methods being the most promising. Despite being critical for drug transport across cancer cell membranes, the interaction of drug-loaded nanoparticles with biological membranes is typically ignored. It is challenging for continuous transport-based modelling to accurately depict the tumor microenvironment in tissue-scale models. Solid tumor differ from patient to patient due to hereditary variables, making them heterogeneous. While there is potential for high-resolution imaging, there are currently few in vivo applications. To accurately forecast the impact of the tumor microenvironment on drug distribution, efforts must be made to create effective multiscale platforms that connect microscopic and macroscopic processes. The need for patient-specific modelling necessitates the use of diagnostic and follow-up information, including medical pictures, clinical data, and biochemical data. Planning a patient's care before and after surgery can be facilitated with personalized simulations. Preliminary predictions may be made using reduced-order models with patient-specific input, but reliable modelling is difficult due to the system's high nonlinearity and the emergence of various drug resistances.

The impact of tumor structure and size on medication delivery to solid tumor can be assessed using a mathematical equation for fluid flow in physiological systems containing tumor (Soltani et al., 2012). To discretize and solve the governing equations with the necessary boundary conditions, the element-based finite volume approach is employed. By looking at the pressure and velocity of the interstitial fluid, which they assume moves along with the drug particles, the researchers gain greater insight into the administration of medication inside solid tumor. By including non-spherical tumor shapes like prolate and oblate ones, this study broadens its focus. Because spherical tumor are more analytically tractable, earlier research has mostly focused on them. Further research is being done on the impact of tissue surface area per unit volume, vascular conductivity, and interstitial hydraulic conductivity on the administration of medication. Cells, the interstitium, and the vasculature—important tissues involved in the transport of drugs to tumor—as well as the lymphatic system, which is in charge of tissue drainage, are also highlighted. Two of the fibres that make up the interstitium, a gel-like area situated between blood vessels and cells, are collagen and glycosaminoglycans. The vasculature is composed of blood vessels of various sizes. To efficiently reach the patient, the medicine must pass through the blood vessel wall, enter the interstitial region, and pierce the cancer cell membrane. However, drug delivery is usually compromised in solid tumor due to inadequate lymphatic drainage, particularly for large molecules like monoclonal antibodies (MAbs), which raise interstitial pressure. The parameter α/R, which denotes the ratio of vascular permeability to hydraulic conductivity, is mostly unaffected by tumor geometry. However, the tumor's shape affects the pressure profile within it. The interstitial fluid pressure (IFP) function, which depends on r, and R, can be obtained by using a solution assuming a spherical profile. IFP will typically resemble vascular pressure (PB) within the tumor interior, with a high-pressure gradient towards the boundary, despite variations in equations fitted to different tumor forms. Shape disparities are significant for lower tumor radii, but over a certain radius, the shape is less important. The study demonstrates that except for smaller spherical or spheroidal tumor, increasing vascular leakiness does not always promote drug convection. Regularly occurring irregular forms in the body have more noticeable impacts. The study shows that maximum flow is more strongly shaped by size than by shape and that vascular permeability cannot allow uniform drug dispersion because of its low diffusivity.

Mathematical modelling can be used to comprehend and improve how anticancer medications are delivered and how they affect tumor cells (Liu et al., 2011). Biological and physiological characteristics that can obstruct drug delivery are highlighted while discussing the intricate mechanisms involved in drug transport and its interaction with tumor. The review looks at various mathematical models that describe the effects of the drug, such as pharmacokinetic and transport-based models as well as empirical or deterministic models. To fully understand the drug delivery process, the authors stress the importance of integrating cellular signal transduction with biophysical components of drug transport. It is explained how computational models are used in medication distribution and uptake, emphasizing the necessity for more advancements in validity and physiological reliability. The geographical distribution of medication concentration is a problem because current models primarily focus on temporal trends and ignore tumor heterogeneity. To effectively forecast geographic variation, diffusion, and convection mechanisms must be included. The prediction of drug concentration at the cellular target site is further complicated by taking into account cellular reactions, such as the synthesis of Pgp or alternate sequestration. In comparison with intravenous injection, specialized drug delivery systems (DDSs) and intratumoral implantation of drug-releasing devices show promise in enhancing medication penetration within tumor tissues. Modelling of tumor response is also covered. Empirical pharmacodynamic (PD) models, which are frequently calibrated with scant experimental data, provide a quantitative but constrained explanation of concentration-effect or effect-time connections. However, because of the tumor microenvironment and nonlinear system behaviour, it is important to exercise caution when using PD models that have been validated through in vitro investigations for clinical applications. An organized understanding of how extracellular and intracellular variables affect medication response is provided by deterministic PD models that contain intricate cellular signalling networks. Although difficult, developing adequate frameworks that cover cellular signalling is necessary for thorough modelling. Creating multiscale system frameworks that integrate medication administration, transport, and cellular signalling is a difficult task. This requires taking into account data at several levels, including blood flow, drug transport, cellular signalling, and overall tumor response. Future modelling efforts should concentrate on combining systems biology descriptions of cellular behaviour and synthesizing insights from specific contributing elements into suitable systems frameworks. To better understand and improve drug delivery and efficacy, prediction platforms must be developed due to the intricacy of drug effects on cells.

6 Personalized Treatment for Cancer Study

Personalized cancer treatment is a useful method to curate personalized therapy for individual patients as it helps target the root cause of the problem faced by the patient. Computational fluid dynamics (CFD) is utilized in designing this personalized treatment by utilizing methods like drug delivery optimization, therapeutic agent transport, treatment response assessment, and radiation therapy planning.

Personalized treatment also known as precision medicine is designed using multiple computational techniques ranging from the unique character study of the patient by studying the genomic analysis, predictive modelling, drug sensitivity and resistance, treatment optimization, and many others. Some techniques and studies completed using computational fluid dynamics have been explained here in brief to understand the applications of CFD properly in designing a personalized treatment for cancer patients.

Drug delivery optimization can be analysed using CFD simulations, the simulations can be employed to strategize the personalized treatment by considering factors like tumor location, size, and vascularization. CFD modelling helps in evaluating the effectiveness of different drug delivery methods and provides insights on drug distribution patterns within the tumor tissue, estimates drug concentrations at the target site, and guides the selection of optimal delivery methods for different patients.

The study of the movement and distribution of various therapeutic drugs inside tumor is done using therapeutic agent transport. Tumor geometry, vascular networks, and fluid dynamics can all be used as various inputs in CFD models to assist predict the diffusion and interaction of medications or nanoparticles inside the microenvironment. The knowledge gained from the spatial and temporal distribution of therapeutic drugs allows one to comprehend the effects of interstitial fluid flow on drug transport and potential barriers to effective drug delivery within tumor.

Treatment response assessment through CFD simulations can aid the assessment of treatment response and guide personalized treatment decisions. By including the patient-specific data, CFD simulations can provide quantitative metrics and spatial information about the treatment response, aiding clinicians in evaluating therapy effectiveness and curating plans potentially suitable to individual patient characteristics.

Radiation therapy planning is widely used for patient-specific treatment using CFD simulations. CFD modelling by simulating the flow of radiation dose within the body can help to predict the distribution of radiation within the tumor and surrounding healthy tissues. These simulations can assist in analysing and studying the optimal beam radiation beam angles, intestines, and durations to maximize tumor irradiation while minimizing damage to healthy tissues. CFD modelling while considering the patient-specific anatomical and physiological factors proves to be a helpful factor in designing the patient-specific therapy.

The relationship between objective assessments of laryngeal airway resistance and clinical respiratory distress was investigated using computational fluid dynamics (CFD) and morphometric analysis. (Hudson et al., 2023). Retrospective CT data of 20 patients with laryngeal cancer and 20 healthy controls were analysed. To evaluate mass flow, wall characteristics, and pressure conditions, CFD models were developed. Convergence research was carried out to identify the ideal voxel size for CFD simulations (Fig. 5). With varied voxel lengths, simulations were run on both normal and critically constricted airways. Regions of various resolutions were created within the discretization. The model was validated using typical airways, and supercomputer network parallelization enabled computation times to be cut from days to hours. Although computational methods have made tremendous progress in several domains, their use in clinical medicine, particularly in otolaryngology, is still in its infancy. The objective of this study was to determine how well computational fluid dynamics (CFD) models could predict airway obstruction in laryngeal cancer. The association between calculated airway resistance and acute obstruction was substantial, demonstrating the potential of CFD analysis for assessing resistance and foretelling airway distress. However, further improvement is required because the current computation times are not practical for clinical use. Despite some drawbacks, including a small sample size and retroactive data collection, the work sheds light on some prospective CFD uses in otolaryngology. Airflow resistance and airway blockage have a substantial correlation in laryngeal cancer patients, according to CFD research, which raises the possibility of a predictive model that incorporates clinical variables, CFD, and morphometric measures.

7 Case Study of CFD to Evaluate the Effectiveness and Precision of Specific Cancer Treatment Methods

7.1 Intra-arterial Chemotherapy

According to the outcomes of the CFD model for intra-arterial chemotherapy, a curved catheter tip may be positioned below and in the direction of the target artery to enhance the rate of anticancer agent distribution in the tumor-feeding artery. The fluid model analysis makes it simpler to interpret the flexibility of the artery walls and catheter movement in varied patient scenarios.

When compared to systemic intravenous chemotherapy, super selective intra-arterial chemotherapy (SSIAC) provides a higher concentration of anticancer drugs into the arteries that nourish the tumor (Kitajima et al., 2020). It is unknown how the medication is delivered between the lingual artery and facial artery (FA) in patients with the linguofacial trunk. In this study, agent dispersion in SSIAC was examined using CFD (Fig. 6). Catheter models and vascular models from two patients' CT scans were used to simulate intra-arterial infusion. The study highlights the need to alter the position per vascular topology by showing that the distribution of the agent is affected by the position of the catheter tip in the linguofacial trunk. For forecasting agent flow in particular patients before therapy, CFD can be a useful technique. The distribution of anticancer drugs during super selective intra-arterial

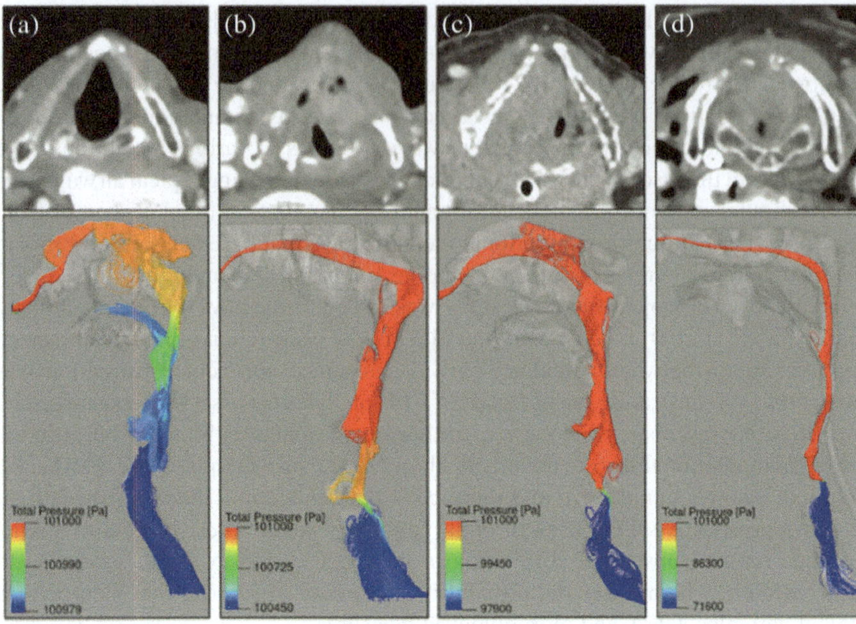

Fig. 5 Computational fluid dynamic study showing pressure distributions from the nostril to the trachea and axial CT images: **a** Advanced transglottic lesion, **b** early stage glottic lesion, obstructions in **c** and **d** are severe and very severe, necessitating urgent (within 48 h) and emergency (immediate) airway surgery, respectively (Hudson et al., 2023)

chemotherapy (SSIAC) for patients with the linguofacial trunk was examined in a recent CFD study. The analysis revealed that the agent's dispersion did not follow the expected blood distribution between the lingual artery (LA) and the facial artery (FA). Instead, it is based on the blood flow field of each patient, which varies due to variations in artery structure. The research highlighted the significance of the blood flow field in a 2 mm radius around the catheter tip for identifying the agent's final resting location. Based on catheter position, the findings suggest employing models to predict agent distribution for certain individuals before treatment.

The mathematical modelling of drug transport during intraperitoneal chemotherapy (IPC) for peritoneal metastases is the main topic of this study. The researchers improved their earlier work by incorporating actual tumor shapes and regionally variable vascular characteristics using a 3D CFD model. DCE-MRI imaging was used to distinguish between tumorous tissues, healthy surrounding tissues, and necrotic zones based on variations in vascular characteristics. A considerable effect on the interstitial pressure profiles within tumor was shown to be caused by uneven geometries and different zones, according to the study. The penetration depths of the drug cisplatin in tumor-specific conditions ranged from 0.32 mm to 0.50 mm. Additionally, the results demonstrated that zones with different vascular properties affected the relationship between tumor size and interstitial fluid pressure

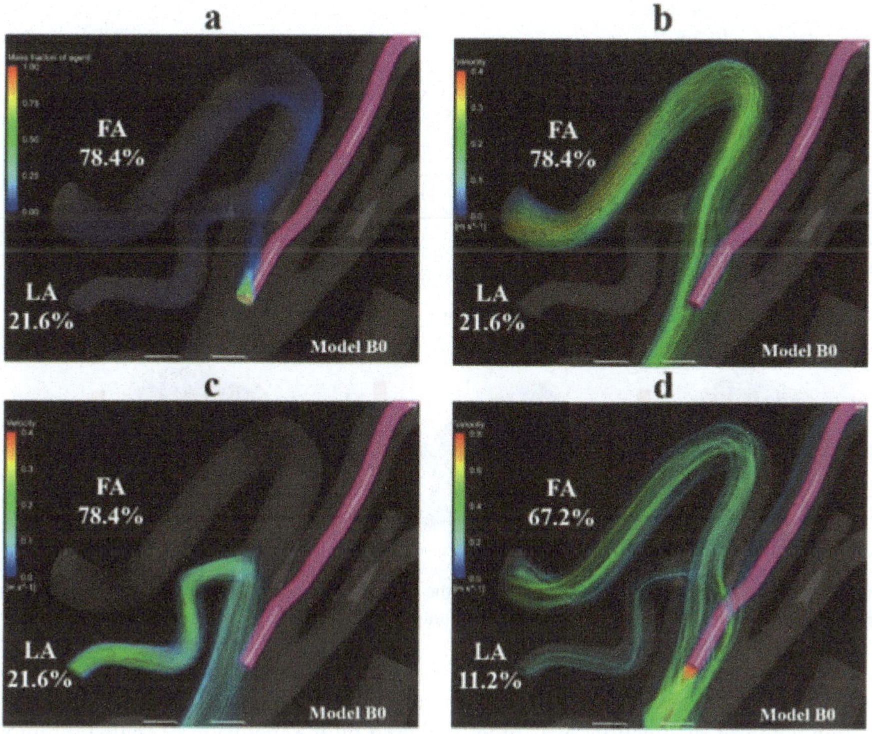

Fig. 6 Streamlines and volume rendering in a vessel model created using patient medical pictures. (a) Volume rendering of the agent mass fraction under the syringe driver's injection condition; (b) streamlines towards the facial artery and (c) streamlines towards the lingual artery (d) streams starting at the catheter tip with the syringe driver's injection state (Kitajima et al., 2020)

(IFP), but a positive correlation was discovered between the percentage of viable tumor tissue and maximum IFP. To better understand the kinetics of drug transport, this study highlights the importance of including uneven tumor geometries and varied vascular zones in CFD models of IPC. Using Mimics software, three alternative tumor geometries were chosen and segmented (Fig. 7). Following smoothing and meshing in 3-Matic, the obtained geometries were imported as STL files into COMSOL Multiphysics. Using a similar technique, other interior zones within the tumor were also obtained. COMSOL was used to construct volume meshes with a constant element size of $1.7 \times 10^{(-4)}$ mm3. Based on earlier explanations by the researchers, the equations for mass transport and interstitial fluid pressure (IFP) build-up were put into practice. To more accurately simulate drug mass transfer during intraperitoneal chemotherapy (IPC), a 3D computational fluid dynamics (CFD) model was developed to include realistic tumor geometries and a variety of vascular characteristics. To distinguish between various vascular characteristics in tumor, healthy tissues, and necrotic zones, DCE-MRI imaging was employed. When locations with various vascular characteristics were taken into account, the study found that the maximal

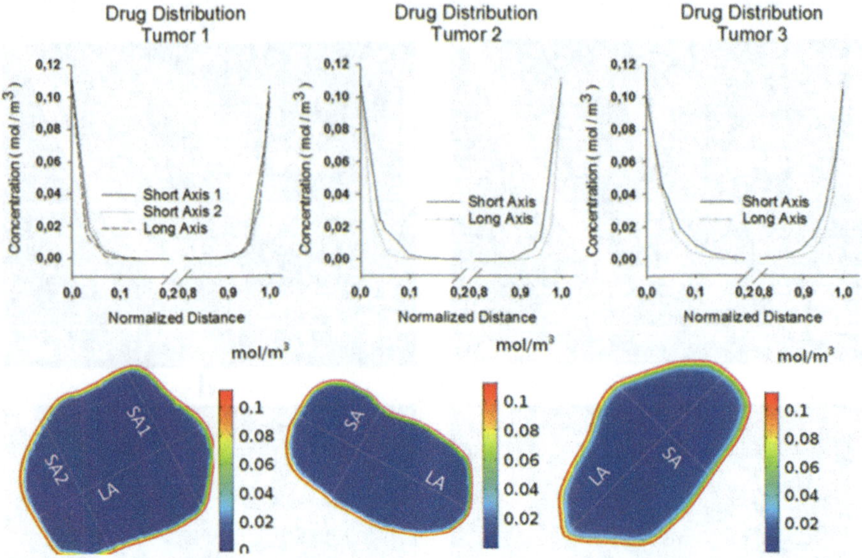

Fig. 7 Figure showing the drug distribution in the various heterogeneous tumor geometries along the axis during intraperitoneal chemotherapy (Steuperaert et al., 2019)

interstitial fluid pressure (IFP) was no longer connected with tumor size. The penetration depth of medications was greatly altered by the inclusion of realistic tumor geometries and necrotic cores. Uneven geometries and various zones greatly influenced the pressure profiles within tumor. The study emphasizes how crucial it is to incorporate both variables into the model to comprehend drug transport during IPC (Steuperaert et al., 2019).

The flow distribution of anticancer drugs into branches of the external carotid artery (ECA) can be examined using CFD to increase the efficacy of intra-arterial chemotherapy (IAC) for mouth cancer (Kitajima et al., 2017). Patient-specific vascular and catheter models were created, and several IAC procedures were simulated. The models were used to look at the distribution of anticancer drugs and the shear stress in the arteries walls. Due to the anticancer drugs' insufficient penetration into the arteries that nourish the tumor, IAC for oral cancer has not proven to be sufficiently effective. When manually supplied during catheterization, anticancer drugs may not effectively reach the artery that feeds the tumor, producing false-positive results. CFD was utilized to analyse the anticancer medication flow dispersion in various IAC approaches to address these issues. Models of the patient's catheters and vessels were created using CT scans, and simulations were done. The study found that placing the catheter tip below and in the general direction of the target artery increased the rate at which anticancer medicines were distributed into the target artery. On the other hand, the distribution rate to various branches varied depending on where the catheter was positioned. Wall shear stress (WSS), which affects endothelial function, was also investigated. The work suggests that catheter insertion might be planned using WSS models to safeguard the tumor-feeding artery. The study has significant limitations, such as the assumption of healthy blood vessels and the small sample size, but despite these, it provides recommendations for enhancing IAC methods for the treatment of oral cancer. To fully understand the distribution of anti-cancer medications in the branches of the external carotid artery, additional studies with more patient data are required. According to computational analysis of IAC, a curved catheter tip can maximize the distribution of anticancer drugs; nevertheless, care should be taken due to vascular elasticity and wall shear stress.

7.2 *Radioembolization (RE)*

This work aims to optimize the radioembolization (RE) method employing yttrium-90 (90Y) microspheres for liver cancer patients by correlating in vivo data with computational modelling. CFD software was utilized to forecast the 3D hemodynamics and microsphere distribution in the hepatic artery during RE using in vivo data from three patients with hepatocellular cancer. The distribution of microspheres in the simulation was compared to the actual distribution as established by 90Y PET/CT imaging. The results showed that there was only a small average difference between the activity distribution based on PET/CT and the activity distribution based on CFD, suggesting that CFD simulations can forecast 90Y-microsphere distribution and possibly optimize the RE procedure for particular patients. The number of 90Y-microspheres that reached each liver area was compared using data from 90Y PET/CT scan treatments and computational models utilizing CFD. The mean activity per volume was calculated from the PET/CT images, and the activity per volume for each region was estimated using the CFD simulation. Three patients with liver cancer participated in this comparison. The distribution of 90Y-microspheres following injection into the hepatic artery has been predicted using a 3D computer model that combines patient-specific data and geographical traits. In vivo testing has proven the model's ability to customize radioembolization treatments (Anton et al., 2021).

CFD modelling is utilized to better comprehend and improve the efficacy of yttrium-90 radioembolization (RE), a treatment for terminal liver cancer. Over the past ten years, computer models of the hemodynamics of the hepatic arteries during RE have been carried out to understand more about the process. As a part of the in vivo validation, microspheres were injected into one cardiac cycle, and the dispersion of the microspheres was extrapolated from segment to segment. Comparisons were made between the predicted microsphere distributions from CFD simulations and the actual distributions discovered using Y-90 PET/CT imaging. The study demonstrated that the CFD model could successfully forecast the microsphere dispersion. The next step is to use CFD models as a pre-RE tool to improve the treatment parameters and forecast the distribution of microspheres in the liver. A computer simulation-based investigation on the hemodynamics of the hepatic arteries following yttrium-90 radioembolization (RE) has produced insightful results. The challenges of blood flow and microsphere transport in tortuous arterial trees, the importance of injection position and velocity near bifurcations, the impact of catheter type on microsphere distribution, and the predictability of segment-to-segment microsphere distribution are just a few of the lessons that could be taken away from this study. Clinicians are recommended to employ patient-specific CFD models with accurate boundary conditions, optimize injection parameters, and consider catheter type and position as a result of these teachings. The goal is to improve the patient's understanding of and response to RE therapy.

In the context of radioembolization therapy for liver cancer, the effect of changing the shape of the hepatic artery on simulation time was also examined in the research (Lertxundi et al., 2021). According to the results, artery truncation can greatly shorten simulation time without materially altering microsphere distribution. This study investigates if the geometry of the hepatic artery can be made simpler to shorten the computational time in radioembolization (RE) simulations. Hepatic artery trees unique to each patient were examined, and a plan was made to cut out pointless branches. The results show that while greatly lowering simulation time, shortened geometries might match baseline simulations quite accurately. It was noted, nonetheless, that the imposed blood flow split did not always match the distribution of the microspheres. The work emphasizes the necessity of including local hemodynamic events during RE in CFD calculations. Although there is still room for improvement, cutting down on computational time can lead to more effective treatment planning. In RE simulations, precise microsphere distribution prediction depends on patient-specific CFD analysis (Fig. 8). Planning for tailored therapy is made easier by shorter geometries, which produce comparable results with a large reduction in simulation time.

A hybrid particle-flow modelling approach for computational fluid dynamics (CFD) analysis in the treatment of liver cancer can be introduced, along with a truncation algorithm to make hepatic arterial trees simpler (Bomberna et al., 2022). The findings suggest that truncating the vascular tree can be used to accurately estimate particle distribution, with the hybrid model outperforming flow distribution alone in terms of accuracy. To validate these results in more patient-specific geometries, additional study is required. The accuracy and computational complexity of four computational methods for calculating particle dispersion in hepatic artery geometry are compared. There is a 27% reduction in overall computational cost when the hepatic artery geometry is simplified from Geometry 1 to Geometry 3. This is because there are fewer mesh elements and convergence occurs more quickly. Consistent trends may be seen in the CPRGs of various hepatic artery geometries, indicating preferred injection sites. When compared to modelling merely flow, truncation has a minimal impact on accuracy, while modelling particles increases accuracy. There are only slight variations in the behaviour of microparticles between catheter injections and the full-complexity particle distribution. In comparison with truncating to Geometry 2, truncating from Geometry 1 to Geometry 3 has a lower accuracy loss and a bigger computational time reduction. Although less crucial for planar injections, explicitly describing the particle dispersion is nevertheless vital for catheter injections.

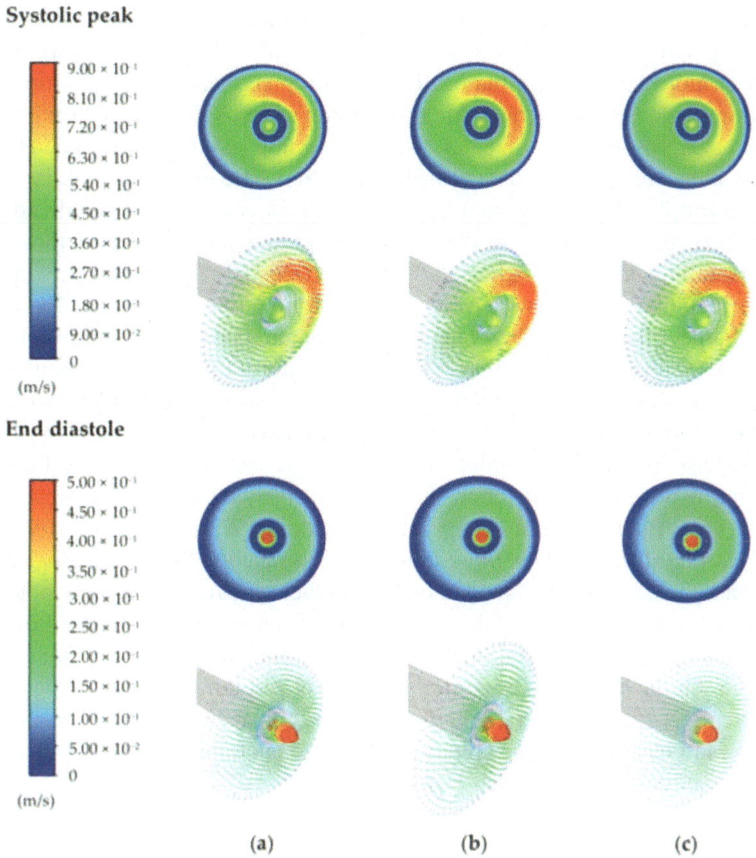

Fig. 8 Velocity magnitude contours and vectors: (a) baseline geometry; (b) upstream truncation (3 cm from the microcatheter-tip location and downstream truncation); and (c) upstream truncation (4 cm from the microcatheter-tip location and downstream truncation) (Lertxundi et al., 2021)

To simplify computations in CFD simulations of trans-arterial radioembolization treating liver cancers, this paper suggests using a hybrid particle-flow model. The paper presents suggestions for future improvements and validation, highlights limitations, and suggests novel techniques.

7.3 Intraperitoneal Chemotherapy

Intraperitoneal (IP) chemotherapy is a method of administering chemotherapy into the abdomen (also known as the peritoneal cavity) rather than a vein. Cancer cells can be treated directly by injecting chemotherapy into the abdomen. This work provided knowledge to optimize treatment parameters by examining the effects of droplet size, flow rate, viscosity, and electrostatic field on the spatial distribution of intraperitoneal aerosolized anticancer medicines (Rahimi-Gorji et al., 2022). The ePIPAC process produced a more uniform distribution of aerosol droplets than PIPAC, according to the study's simulations of the geographic distribution of aerosol droplets during the two procedures. Aerosol droplet size has a substantial impact on how they deposit, with larger droplets being more gravitationally impacted. The most uniform distribution is seen in droplets between 1 and 5 m. Higher flow rates during nebulization led to more aerosol particles being deposited in region 1 in CFD simulations. Large droplets and low flow rates indicated deposition, mostly in the nebulizer's opposite direction. Smaller droplet sizes were seen when flow rates increased. The aerosol droplet diameter and liquid flow rate affect the amount of inertial impaction during deposition. Increasing the flow rate in PIPAC procedures to 0.6 mL/s can improve the homogeneity of droplet distribution. The cone angle and aerosol droplet nebulization were studied in an experiment. The cone angle increases together with the flow rate until it reaches its maximum value. The cone angle and flow rate were discovered to have an exponential connection. Icodextrin had dynamic viscosity readings of 1.88 mPa s (4%) and 2.24 mPa s (7.5%) at various doses. With lower-viscosity liquids, aerosol droplet distribution is more homogeneous, according to CFD models. The electrostatic field was investigated in the box model at different electrical potentials, and it was found that it was more powerful at higher potentials. Aerosol dispersal improved with higher potential, with 6.5 kV being the ideal potential. This research sheds light on the best ways to administer aerosolized intraperitoneal drugs for the treatment of peritoneal cancer, highlighting the significance of droplet size, flow rate, viscosity, and electrostatic precipitation. Realistic anatomical models and droplet-tissue interactions will be studied further.

7.4 Modelling of Endovascular Chemofilter Device

The Chemofilter is an endovascular filtering device that removes chemotherapy medicines from blood to increase treatment effectiveness. It was designed using a two-scale CFD modelling technique. Using a multiscale approach and CFD simulations, the Chemofilter device was investigated to optimize its design for efficient drug binding and little flow obstruction (Fig. 9). The hemodynamics of the Chemofilter device were investigated using multiscale CFD simulations (Maani et al., 2018). The Chemofilter membrane design was optimized using multiscale CFD simulations, taking into account elements like microcell size, layer count, tip angle, and eliminating gaps for greater filtration effectiveness. Taking into account variables including microcell size, membrane shape, and lattice orientation, the best membrane designs were found to reduce pressure drop while optimizing drug binding and flow residence time. The 40° tip angle, four-sector basket membrane was designed using the findings. While taking into account restrictions such as vascular compliance and

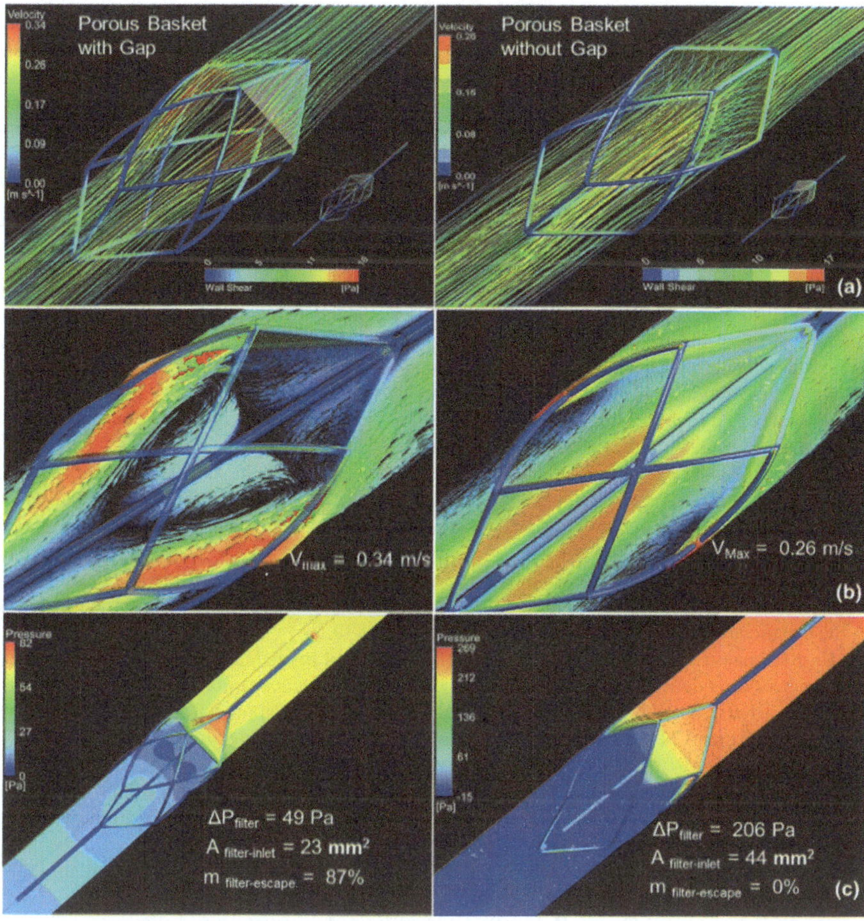

Fig. 9 Distribution of shear stress, velocity vectors, and pressure profile of flow streamlines going through the Chemofilter (Maani et al., 2018)

device apposition, high-dose heparin can be supplied to reduce thrombus development and optimize the Chemofilter device. Modelling chemical binding and assessing membrane flexibility are other developments.

References

Antón, R., Antoñana, J., Aramburu, J., Ezponda, A., Prieto, E., Andonegui, A., ... & Rodríguez-Fraile, M. (2021). A proof-of-concept study of the in-vivo validation of a computational fluid dynamics model of personalized radioembolization. *Scientific Reports*, *11*(1), 3895

Bomberna, T., Vermijs, S., Lejoly, M., Verslype, C., Bonne, L., Maleux, G., & Debbaut, C. (2022). A hybrid particle-flow CFD modelling approach in truncated hepatic arterial trees for liver radioembolization: A patient-specific case study. *Frontiers in Bioengineering and Biotechnology, 10*, 914979.

Hudson, T. J., Ait Oubahou, R., Mongeau, L., & Kost, K. (2023). Airway Resistance and Respiratory Distress in Laryngeal Cancer: A Computational Fluid Dynamics Study. *The Laryngoscope*.

Kashkooli, F. M., Soltani, M., & Hamedi, M. H. (2020). Drug delivery to solid tumors with heterogeneous microvascular networks: Novel insights from image-based numerical modelling. *European Journal of Pharmaceutical Sciences, 151*, 105399.

Kitajima, H., Iwai, T., Yajima, Y., & Mitsudo, K. (2020). Computational Fluid Dynamics Study of Superselective Intra-arterial Chemotherapy for Oral Cancer: Flow Simulation of Anticancer Agent in the Linguofacial Trunk. *Applied Sciences, 10*(21), 7496.

Kitajima, H., Oshima, M., Iwai, T., Ohhara, Y., Yajima, Y., Mitsudo, K., & Tohnai, I. (2017). Computational fluid dynamics study of intra-arterial chemotherapy for oral cancer. *BioMedical Engineering OnLine, 16*(1), 1–26.

Koumoutsakos, P., Pivkin, I., & Milde, F. (2013). The fluid mechanics of cancer and its therapy. Annual review of fluid mechanics, 45, 325–355.

Kuttler, A., Dimke, T., Kern, S., Helmlinger, G., Stanski, D., & Finelli, L. A. (2010). Understanding pharmacokinetics using realistic computational models of fluid dynamics: Biosimulation of drug distribution within the CSF space for intrathecal drugs. *Journal of Pharmacokinetics and Pharmacodynamics, 37*, 629–644.

Lertxundi, U., Aramburu, J., Ortega, J., Rodríguez-Fraile, M., Sangro, B., Bilbao, J. I., & Antón, R. (2021). CFD simulations of radioembolization: A proof-of-concept study on the impact of the hepatic artery tree truncation. *Mathematics, 9*(8), 839.

Liu, C., Krishnan, J., Stebbing, J., & Xu, X. Y. (2011). Use of mathematical models to understand anticancer drug delivery and its effect on solid tumors. *Pharmacogenomics, 12*(9), 1337–1348.

Maani, N., Hetts, S. W., & Rayz, V. L. (2018). A two-scale approach for CFD modelling of endovascular Chemofilter device. *Biomechanics and Modelling in Mechanobiology, 17*, 1811–1820.

Rahimi-Gorji, M., Debbaut, C., Ghorbaniasl, G., Cosyns, S., Willaert, W., & Ceelen, W. (2022). Optimization of intraperitoneal aerosolized drug delivery using computational fluid dynamics (CFD) modelling. *Scientific Reports, 12*(1), 6305.

Rajput, S., Sharma, P. K., & Malviya, R. (2023). Fluid Mechanics in Circulating Tumour Cells: Role in Metastasis and Treatment Strategies. *Medicine in Drug Discovery*, 100158.

Shojaee, P., & Niroomand-Oscuii, H. (2019). CFD analysis of drug uptake and elimination through vascularized cancerous tissue. *Biomedical Physics & Engineering Express, 5*(3), 035032.

Soltani, M., & Chen, P. (2011). Numerical modelling of fluid flow in solid tumors. *PLoS ONE, 6*(6), e20344.

Soltani, M., & Chen, P. (2012). Effect of tumor shape and size on drug delivery to solid tumors. *Journal of Biological Engineering, 6*, 1–15.

Steuperaert, M., Debbaut, C., Carlier, C., De Wever, O., Descamps, B., Vanhove, C., ... & Segers, P. (2019). A 3D CFD model of the interstitial fluid pressure and drug distribution in heterogeneous tumor nodules during intraperitoneal chemotherapy. *Drug delivery, 26*(1), 404–415.

Vulović, A., Šušteršič, T., Cvijić, S., Ibrić, S., & Filipović, N. (2018). Coupled in silico platform: Computational fluid dynamics (CFD) and physiologically-based pharmacokinetic (PBPK) modelling. *European Journal of Pharmaceutical Sciences, 113*, 171–184.

Wu, C., Hormuth, D. A., Oliver, T. A., Pineda, F., Lorenzo, G., Karczmar, G. S., ... & Yankeelov, T. E. (2020). Patient-specific characterization of breast cancer hemodynamics using image-guided computational fluid dynamics. *IEEE transactions on medical imaging, 39*(9), 2760-2771.

Zhan, W., Alamer, M., & Xu, X. Y. (2018). Computational modelling of drug delivery to solid tumour: Understanding the interplay between chemotherapeutics and biological system for optimised delivery systems. *Advanced Drug Delivery Reviews, 132*, 81–103.

Chapter 7
Computational Fluid Dynamics for Modelling and Simulation of Drug Delivery

1 Introduction

The amount of information we have of both the biological and physical processes involved in medication administration during the past century has greatly increased, as has our ability to manage drug delivery systems. By using modelling and simulation to take advantage of the exponential increase in computing power, it is possible to learn new, complementary knowledge about the methods and components used in medication delivery. Computational fluid dynamics (CFD) is crucial in drug delivery systems because it helps to clarify the key drug release mechanisms, allowing for the methodical creation of novel pharmaceutical products as opposed to a trial-and-error method. Experimental data can be used in a variety of real world contexts, as is demonstrated. Traditional research methodologies have been shown to be insufficient, and the rising cost of implementation is another effect of this expanding complexity. The concepts of mathematical modelling and computational simulations related to drug distribution are discussed in this chapter. The effects of computer simulation and modelling on research into the administration of drugs via the nervous system, ocular system, transdermal system and respiratory system are discussed.

2 Background

Drug delivery is the process of administering therapeutic substances such as medications or drugs to achieve a desired therapeutic effect in the body. The goal of the drug delivery system is to improve drug effectiveness and safety by managing drug release, targeting specific regions of the body, and optimizing pharmacokinetic and pharmacodynamic characteristics. Oral intake, injections, and topical treatments are examples of classic medication delivery modalities. However, advances in pharmaceutical research and technology have resulted in the development of a variety of novel drug

S. E. Jujjavarapu et al., *Computational Fluid Dynamics Applications in Bio and Biomedical Processes*, https://doi.org/10.1007/978-981-99-7129-9_7

delivery systems. Controlled release systems (release of drug progressively over an extended period maintaining therapeutic levels throughout the body and reducing frequent dosing), targeted drug delivery (the delivery of drugs to particular sites in the body like tumor or inflamed tissues whereas minimizing systemic exposure and side effects) are the various and notable approaches for drug delivery. Targeting is accomplished through active processes or passive accumulation of specific tissues), nanotechnology-based drug delivery (nanoscale delivery employs nanoparticles or nanocarriers for drug transportation), inhalation delivery (inhalers and nebulizers are used to deliver drugs directly into the airways, rendering them appropriate for treating respiratory conditions), gene and cell therapy (advanced therapies involving the delivery of genetic materials or cells). Drug delivery is a field that is always expanding due to continuing study and technological improvements. The ultimate goal is to increase pharmacological efficacy, compliance among patients, and overall therapeutic results for side effect reduction and treatment regimen optimization.

CFD is a branch of fluid dynamics that uses numerical methods and algorithms to solve and analyse fluid flow and heat transfer problems. It aids engineers and scientists to simulate and predict the behaviour of fluids, such as gases and liquids along with interaction between fluids and solid structures. The key aspects of CFD include mathematical modelling, discretization, solution methods, boundary conditions, visualization, and post-processing. CFD is widely used in industries including aerospace, automotive, energy, chemical, and biomedical engineering. This helps engineers optimize designs, improve efficiency, and reduce costs by simulation of fluid flow phenomena under different operating conditions without any need for physical prototypes. CFD has been an indispensable tool for understanding and predicting fluid behaviour and optimizing engineering processes. CFD has been of use in the field of drug delivery as it provides the opportunity to study the crucial processes computationally. Some of these processes are inhalation drug delivery, nasal drug delivery, transdermal drug delivery, drug eluting stents, micro and nanoparticle drug delivery, and biomedical implants and drug delivery devices. Inhalation drug delivery is employed for the study and optimization of inhaler devices such as metered-dose inhalers (MDIs), dry powder inhalers (DPIs), and nebulizers. CFD simulations are used to analyse the airflow patterns, particle deposition, and drug distribution within the respiratory system. It helps in understanding the drug delivery efficiency, identification of regions of high or low drug deposition, and the optimization of the design parameters for inhalers. CFD is also used for investigation of the transportation and deposition of drug particles of the nasal cavity. This aids in the understanding of complex airflow patterns, particle deposition efficiency, and drug distribution on the nasal mucosa. The simulations are in optimization of nasal drug delivery devices and its formulations for improving drug absorption and therapeutic efficacy. Transdermal drug delivery, another application of CFD in drug delivery, assists in the development of transdermal drug delivery systems. This helps in the understanding of diffusion of drugs through the skin and optimizing factors such as drug concentration, formulation viscosity, and skin permeability. CFD simulations aid in the prediction of drug release rates, studying the effect of different patch designs, and optimization of drug delivery parameters for efficient transdermal delivery. CFD are utilized to study the drug release and distribution from drug eluting stents (DES) which are used in cardiovascular interventions. This helps in the optimization of the design of DES, understanding drug release kinetics, and prediction of the drug concentration gradients within the wall of the artery. This simulation assists in the evaluation of the effectiveness of the DES for preventing restenosis and optimization of drug dosing strategies. Micro and nanoparticle drug delivery is another application for drug delivery that is used to investigate the behaviour of micro and nanoparticles within the drug delivery systems. This helps to understand the particle–particle and particle–fluid interactions, prediction of particle trajectories, and analysis of deposition patterns for different organs and tissues. These simulations help in the designing and optimization of drug delivery systems that employ particles like liposomes, microspheres, or nanoparticles. Biomedical implants and drug delivery is a branch of the application of CFD in drug delivery where the analysis of flow characteristics and drug distribution within the implants or drug delivery devices. This helps in optimization of implant designs, studying drug release profiles, and prediction of drug concentration gradients in the target tissue. CFD simulations help in the evaluation of performance of implants and optimization of drug delivery parameters for enhancement of therapeutic outcomes. CFD simulations are a very useful tool for gaining insights in the transport, deposition, and drug distribution in various drug delivery devices.

Here, in this chapter, we have discussed some common yet complex studies of drug deliveries carried out by scientists across the globe. Drug delivery research on respiratory system, nervous system, ocular system, and transdermal systems are discussed in detail.

3 Case Study for Drug Delivery Application in CFD

3.1 Respiratory System

Respiratory system is a complex network of organs and tissues which facilitate exchange of oxygen and carbon dioxide between the body and environment. This is responsible for the respiration process that involves the oxygen intake and carbon dioxide elimination. Drug delivery in the respiratory system refers to the administration of medications directly into the lungs or airways which treat respiratory conditions to deliver systemic drugs. This can be achieved through different routes that include inhalation, nebulization, and intranasal administration. Application of CFD is a numerical simulation technique which can be used to study the behaviour of fluids, including airflow in the respiratory system. These simulations provide insights to drug delivery in the respiratory system by analysis of the dispersion and deposition of inhaled medications. Airflow modelling, particle dispersion, deposition analysis, inhaler design and optimization, and patient-specific simulations are the studies performed for respiratory system drug delivery analysis.

A unique nasal medication delivery device that enhances drug dispersion while minimizing lung deposition is being developed by OptiNose AS (Kleven et al., 2005). The design and operation of the bidirectional approach are optimized using CFD simulations, which eliminates the need for expensive and time-consuming laboratory tests. The strategy for creating a surface grid of the nasal cavity is described in the study, along with testing methods and validation against successful physical experiments. The bidirectional medication delivery system created by OptiNose AS was subjected to simulations using CFD, the results of which are presented in this publication (Fig. 1). The simulations show the nasal cavity's flow patterns and particle distribution. The findings demonstrate that more particles settle in the nasal cavity when entering the left nostril as opposed to the right. The models also examine the distribution of gas velocities, with the highest velocities seen in the narrowest airway passages. According to the study, particle deposition rates rise as air velocity and particle size rise. The comparison analysis is concentrated on airflow simulations at a rate of 6 l/min. Inertia forces cause larger particles to follow the airstream less closely when comparing particle sizes. The outcomes of the simulations offer insightful information for improving the efficiency of the drug delivery system. Fluent simulations and gamma-scintigraphic studies are compared, and the results are encouraging while also pointing out certain shortcomings. In comparison with the findings from gamma-scintigraphy, Fluent simulations overestimate the deposition in the nasal cavity for particles with a uniform size of 3.5 m. Similar inconsistencies are seen for particles that are uniformly 10 m in size. Errors can be attributed to things like the intricacy of the nasal airway, individual differences, and the Fluent model's presumptions. The inconsistencies result from the Fluent simulations' assumption that all particles are of identical size while the gamma-scintigraphy studies take into account particle size distributions. Particle deposition may also be impacted by the Fluent model's assumption of laminar flow. Future studies will involve additional evaluation and advancements, including the incorporation of turbulence models. This study illustrates the capability of CT scans to produce accurate nose geometry models for CFD simulations. In the process of developing new medical devices, CFD calculations may replace time-consuming and expensive clinical trials and studies. Additional testing against physical experiments is required. The nasal attenuation factor, the Fluent model, and grid refinement must all be improved. The accuracy of simulations can be improved by using turbulence models and particle size distributions. It would be fascinating to explore the implications of anatomical variations by developing grids for various nasal morphologies.

Researchers also studied how tracheal stenosis affected respiratory processes (Fig. 2). The bronchial tree was modelled in three dimensions using CT scans and patient-specific computer simulations (Taherian et al., 2015). The models showed that the stenosis had an impact on the airflow and particle distribution. The impact of stenosis on flow parameters and particle deposition was shown by comparisons before and after the intervention. For larger particles, the results demonstrated increased particle deposition, but for smaller particles, there was no discernible difference. After the stenosis, the pressure drop was also noticeably enhanced. The computer simulations' experimental validation revealed good agreement. The use of idealized

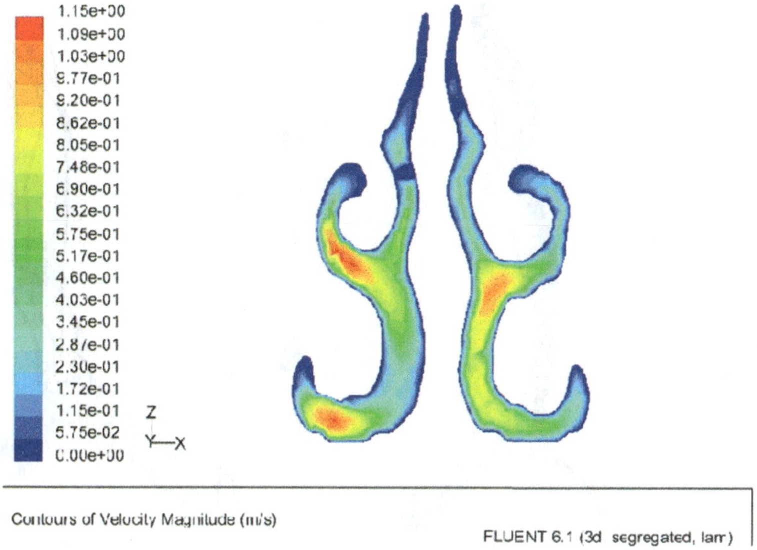

Fig. 1 Computational fluid dynamics of the bidirectional nasal drug delivery distribution with velocity in a nasal cavity cross-section of 6 l/min airflow (Kleven et al., 2005)

models in numerous research to investigate tracheal stenosis has brought attention to the significance of geometry in particle deposition and drug delivery simulations. This study focuses on a 6- to 8-generation CT-based model of a patient's lung with tracheal stenosis. In earlier research, boundary conditions, such as velocity and pressure, have been allocated in various ways. The K-SST turbulence model is used in the current work together with mesh independence analysis to produce precise predictions. Particle transport and deposition simulation are performed using Lagrangian phase modelling. The mucus layer in the trachea is taken into account using a stick wall boundary condition. The study looks into how tracheal stenosis affects the respiratory system's flow traits, pressure distribution, and particle deposition. The findings show zones of flow acceleration and recirculation during inhalation downstream of the stenosis. At the site of the stenosis, pressure decreases, and maximum wall shear stress is seen. A 3D printed model used for experimental validation reveals good agreement with the results of CFD. Stenosis has an impact on particle deposition, with larger particles depositing more heavily and depositing more heavily in particular generations of the lung. The findings have implications for surgical planning and drug delivery optimization. In this study, the effects of tracheal constriction on particle deposition and airflow in a patient-specific respiratory system are examined using CFD. Significant particle deposition occurs upstream and downstream of the obstacle when there is stenosis, along with turbulent flow, secondary flow, and flow separation. The results highlight how crucial airway geometry, particle size, breathing rate, and lung illness are for comprehending particle deposition and maximizing the

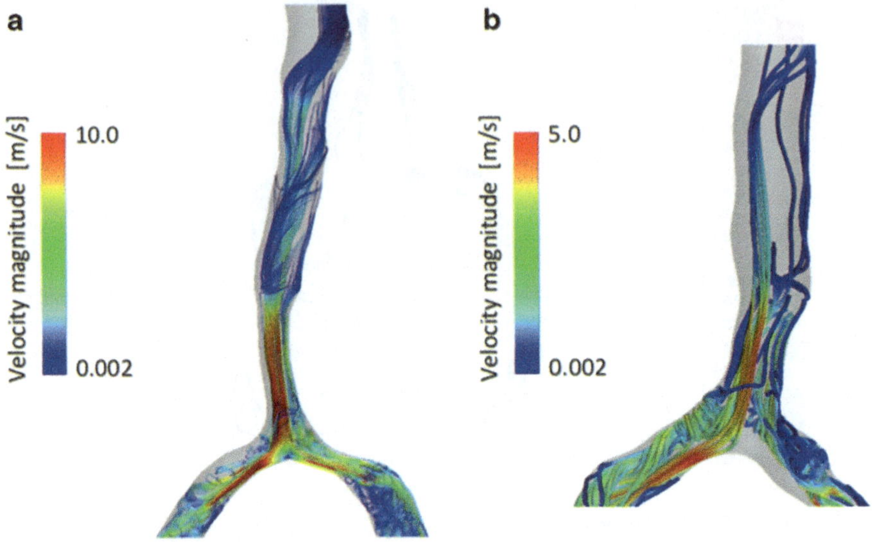

Fig. 2 Effect of velocity magnitude on tracheal stenosis on local flow characteristics: **a** preoperative and **b** post-operative findings (Morita et al., 2022)

efficacy of inhaled medications. To study additional components including the mucus layer and clearance mechanisms, more research is needed.

The benefits of intranasal drug delivery are examined using research, focusing on the treatment of COVID-19 and conditions of the brain and lungs (Tiwari et al., 2022). It shows the significance of CFD in the design and development of effective intranasal medication delivery devices. CFD models can optimize medication deposition in the respiratory tract and nasal airways, increasing the efficiency of drug administration. CFD has been used in numerous research to improve nasal medication delivery systems. Researchers used CFD simulations to increase efficiency and lower expenses, combined with the ECG technique improved lung deposition, utilized Eulerian and Lagrangian methods to study the deposition of nanoparticles. Effective aerosol distribution from nose to lung was accomplished. According to published studies, scientists have mostly employed CT scan data to examine how medication particles are distributed and deposited in the nasal cavity and respiratory system. Complete pulmonary models have been used in several investigations, to forecast drug deposition and distribution. Few articles have discussed altering inhaler designs, although it is thought that CFD in combination with optimization methods can improve the effectiveness of nasal inhalers. According to the research, nasal inhalers have two different types of airflow patterns: axial flow and whirling motion. To ensure that medications enter the lungs deeply while causing the least amount of device loss, swirl airflow at the nasal entrance is preferred.

There are prospective improvements in drug delivery methods as well as the use of CFD in simulating respiratory drug delivery (Longest et al., 2019). The use of regional airway models and CFD to examine aerosol deposition in various respiratory tract regions and how it affects drug absorption is covered in this sample. It presents research on oropharyngeal deposits, laryngeal jet patterns, and alveolar geometry, highlighting how crucial it is to precisely characterize these organs to develop efficient medication delivery methods. Also included are deep inhalation and breath-hold strategies for aerosol administration. The comparison of in vitro experimental data and CFD predictions for pharmaceutical aerosol deposition is covered in this excerpt. There have been studies on several inhaler types, including pressurized metered-dose inhalers (pMDI), soft mist inhalers (SMI), and dry powder inhalers (DPI). Drug deposition in the inhaler alone, upper airways, including the mouth-throat (MT) region, and lower airway bifurcation regions have all been successfully compared quantitatively. To accurately estimate the deposition, variables including turbulent jets, aerosol spray momentum, and droplet evaporation are taken into account. Absolute difference errors are often 5% or less, whereas relative mistakes are typically 10% or less. Due to the fluctuating flow conditions, length scales, lung region compliance, and computational constraints creating numerical models of the entire respiratory system are complex. To address these issues, two strategies have been developed: the whole-lung airway model (WLAM) and the stochastic individual path (SIP) method. These techniques segment the airways into separate sections and simulate flow and particle deposition using anatomically correct geometries. Hybrid 3D/1D techniques and volume-filling branching tree algorithms are two further noteworthy findings. These models shed light on respiratory system aerosol deposition and airflow patterns. New respiratory drug delivery techniques have been developed with the help of CFD models and in vitro tests. Controlled condensational growth is one technique, whereby tiny particles or droplets are inhaled and expand in size within the airways to improve lung deposition. CFD models showed a considerable increase in aerosol size and a depositional loss of less than 1% from the mouth to the throat (MT). Trans-nasal aerosol administration is a different tactic that can be improved by utilizing in vitro tests and CFD simulations. Controlled condensational development and trans-nasal aerosol injection together resulted in high lung deposition and little nasal depositional loss. The effectiveness of aerosol distribution and targeted medicine delivery are improved by these discoveries. Due to their considerable unpredictability and unwillingness to cooperate, aerosol distribution to newborns and children is complex. This problem has been addressed using in vitro investigations and CFD simulations. Studies have examined new delivery methods, devised devices for better medication delivery, and assessed nasal aerosol deposition. For standard aerosol sizes, CFD simulations have demonstrated substantial nose deposition and low lung delivery efficiency. Streamlined Y-connectors have been created to greatly increase lung deposition by lowering deposition losses during mechanical breathing. For improved aerosol administration to newborns with intubations, controlled condensational growth techniques have also been investigated. When studying respiratory flow and aerosol deposition in lung disorders such as cystic fibrosis (CF), chronic obstructive pulmonary disease (COPD), bronchial tumor, and asthma, CFD models have been used. Studies on COPD have primarily examined how bronchodilators affect patients individually using CT scans and CFD models.

Predicting aerosol deposition in CF lungs for antibiotic delivery is one of the emerging applications of CFD, where studies have shown changes in flow and deposition patterns between healthy and CF lungs. These findings offer information on how to make lung disease treatment plans more effective. Researchers are looking into combining computational fluid dynamics (CFD) models with mechanistic simulations of dissolution, absorption, and clearance (DAC), followed by pharmacokinetic (PK) or physiologically based pharmacokinetic (PB-PK) models to accurately predict the distribution of inhaled medications throughout the body. This method combines PK models to evaluate drug concentrations in plasma with CFD predictions of nasal drug uptake. Validation studies have shown a correlation between nasal deposition and plasma drug concentration, however, to achieve bioequivalence in drug delivery, it is important to carefully evaluate elements such as patient-use parameters.

Using CFD, it is even possible to study various drug delivery systems' efficacy in treating respiratory conditions brought on by air pollution (Tiwari et al., 2021). The deposition of medication particles of various sizes was examined using an accurate model of the human respiratory tract (Fig. 3). The findings highlighted the significance of tiny particles in dry powder inhalers by demonstrating that smaller particles were more efficient at penetrating the deeper portions of the lungs. The study examines the effects of three distinct peak inhalation flow rates (60, 30, and 15 LPM) and three different particle sizes (1, 5, and 10 m) on the effectiveness of deposition. There is an increased need for efficient inhalation technology due to the rising prevalence of respiratory illnesses, which is fuelled by air pollution and healthcare costs. Due to environmental considerations, dry powder inhalers (DPIs) are favoured over pressurized metered-dose inhalers (pMDIs). This study uses realistic inhalation profiles through a DPI to examine the distribution patterns of medication particles of various sizes in the human respiratory tract. The findings highlight the significance of particle size and inhalation velocity in the administration of drugs, with smaller particles having a greater ability to enter the distal bronchi. To forecast drug deposition in human airways, the research uses a variety of models and empirical connections. To evaluate the deposition mechanisms, variables including the Stokes number, Froude number, particle Reynolds number, and flow Reynolds number are used. The Froude number shows that inertial impaction predominates over sedimentation. Impaction probability is predicted using a Chan and Lippmann empirical relationship. The results highlight the relationship between flow rate and impaction likelihood and deposition efficiency, with higher flow rates promoting greater deposition in the upper airways and lower flow rates promoting particle reach in the distal region. The study's main conclusions are as follows: the use of a CT scan-based HRT model with DPI inhalation profiles to provide realistic insights into flow and particle deposition in the respiratory system; greater deposition at the beginning of inhalation and decreased deposition with lower inhalation rates; particle size influencing deposition, with smaller particles reaching the distal bronchi more effectively; and the need for further research to determine the ideal flow rate for m.

The effect of a tracheal tumor on airflow patterns and aerosol-drug deposition in the respiratory system can also be examined (Srivastav et al., 2014). Using CT scan data, a realistic model of the human airway with a multifocal tumor was created. The

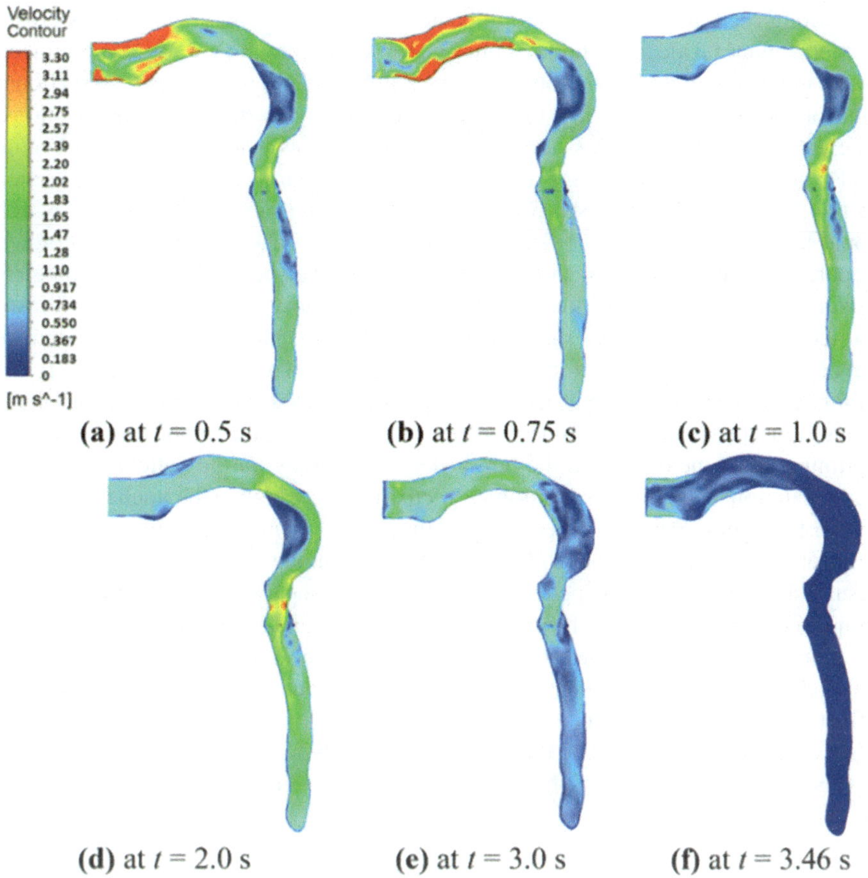

Fig. 3 Computational evaluation of velocity profiles of drug distribution at various times from the oral cavity to the first bifurcation junction in the human respiratory tract under realistic inhalation (Tiwari et al., 2021)

flow disturbances surrounding the tumor and downstream are revealed by numerical simulations, which shed light on the wall shear stress and aerosol-drug deposition. This study contributes to the development of efficient targeted drug delivery systems and our understanding of the consequences of tumor. A tumor in the trachea's posterior wall causes an obstruction that decreases the cross-sectional area and speeds up flow. Different vertical parts of the airway's velocity contours demonstrate how the tumor changes the direction of airflow. Because of the smaller cross-sectional area, the tumor is more pronounced in the downstream zone, where the velocity increases quickly. For cases with and without a tumor, the velocity patterns are comparable in the upstream region, but they alter quickly in the downstream. Around the tumor, the air velocity is at its highest towards the anterior wall, and there is a high-velocity zone close to the posterior wall that could damage the wall. Greater flow disruptions, generating recirculation zones, are indicated by velocity contours at various cross-sectional planes. Due to the tumor's presence, counter-rotating vortices are produced in the downstream area. The tumor blocks a little under half of the cross-sectional area of the airway, according to the average velocity at different cross-sections. The greatest wall shear stress around the tumor remains below the threshold value under normal breathing settings, indicating no wall harm. However, the presence of the tumor greatly increases wall shear stress during laboured breathing, which could lead to epithelial tissue deterioration. Cross portions on the tumor are noted as potential wall damage risk zones. This knowledge can be used by pulmonologists to diagnose, prognosticate, and treat probable respiratory tract wall injury. Aerosol particle size, inspiratory flow rate, and tumor size all affect how effectively medicines are deposited on respiratory tract tumor. Increased deposition efficiency is the result of higher flow rates and larger particle sizes. A particle size exceeding 5 microns is advised for turbulent flow regimes for better deposition on the tumor. Delivering the optimum medicine dosage to the tumor depends heavily on the patient's expertise and inhalation technique. In addition, limited drug deposition in the trachea occurs at higher flow rates, enhancing medication delivery to the tumor. Airflow in the human respiratory system with and without a tumor was numerically simulated, and the results showed altered flow patterns and higher wall shear stress downstream of the tumor. The study provided information for identifying and estimating airway damage by pinpointing the likely site and extent of wall injury. It has been discovered that variables including aerosol size, inhalation flow rate, and patient expertise affect the effectiveness of drug administration. Despite the lack of clinical data, pulmonologists' experiences can validate the findings. The study has implications for bettering disease prognosis and inhaler-based drug delivery systems.

3.2 *Nervous System*

Nervous system is another complex yet important network of cells, tissues, and organs which coordinate and regulate the activities of the body. This is responsible for receiving and transmission of signals between different parts of the body and brain to allow perceiving, processing, and responding capabilities to the world. Drug delivery in the nervous system in the methods and techniques used for the transport of therapeutic drugs or agents specific to the regions or cells within the Central Nervous System or Peripheral Nervous System. The blood–brain barrier poses a challenge in the drug delivery in the CNS as this regulates the passage of substances from the bloodstream into the brain tissue. However, the advancements are still under progress to overcome these concerns actively. CFD is used to simulate and study the drug delivery systems in the nervous system. This is a numerical technique that allows for modelling and analysis of fluid flow associated transport phenomena. The context of drug delivery can provide valuable insights into the distribution and behaviour of drugs within the nervous system aiding the design and optimization of drug delivery systems. CFD is applied in the nervous system as modelling drug dispersion, understanding transport mechanisms, optimization of drug delivery systems, and evaluating drug clearance. The combination of CFD with experimental studies and clinical observations for comprehensive understanding.

In-vitro models and CFD were utilized to evaluate location, volume, pace and physiological aspects in a study to examine medication distribution in the CNS (Khani et al., 2022). The location, volume, pace, and physiological aspects of the injection were evaluated. Compared to lumbar puncture, intracerebroventricular (ICV) and cisterna magna (CM) injections provided greater drug distribution. Drug exposure was improved by raising the volume and including a flush. Factors related to the heart and lungs also affected how drugs disseminate. The results recommend streamlining delivery procedures for focused therapy. In this study, the influence of different factors on the distribution of a tracer in the cerebrospinal fluid (CSF) was examined using a subject-specific multiphase CFD simulation. The variables included catheter position, physiological variables (deep breathing and stroke volume), and injection parameters (bolus volume, bolus rate, and CSF flush). To concentrate on short-term tracer movement inside the CSF, the simulations were run for three hours. Utilizing an in vitro model unique to the subject, the findings were confirmed. It compares CFD results with in vitro data and displays the spatial–temporal distribution of the tracer for lumbar puncture (LP), cisterna magna (CM), and intracerebroventricular (ICV) injections. In contrast to LP, CM and ICV injections demonstrated quicker drug transport to the brain. The steady-streaming theory as the primary driving mechanism for CSF medication administration was corroborated by the in vitro model. For LP, there was a moderate linear association between numerical and in vitro data; for CM and ICV injections, there was a stronger correlation. The tracer dispersion in the CSF system was greatly impacted by the needle's placement. The area under the curve (AUC), maximum concentration (Cmax), and time to maximum concentration (Tmax) of the tracer at various locations along the cerebrospinal fluid (CSF) neuroaxis after three hours are shown, as well as the effects of various injection and physiological parameters. In comparison with LP, CM and ICV injections dramatically increased AUC and increased tracer exposure to the skull. However, the spinal SAS of LP had a larger Cmax. It reveals that the cervical region saw the maximum tracer exposure following the CM injection. Although optimizing injection protocols is needed, intrathecal medication administration is important for both pain relief and CNS therapies. The CSF system was evaluated for medication administration in this work using CFD and in vitro verification. The results show how sensitive medication delivery is to different variables and show how fluid physics-based computational modelling may be used to optimize drug delivery settings. In CFD modelling for biofluid simulations, the selection of numerical solvers and settings might produce varying outcomes. There are currently only invasive or reductionist approaches for measuring medication distribution in CSF. A comparable in vitro model that demonstrated a significant correlation and agreement with experimental results was utilized to validate the CFD model. Additionally, comprehensive numerical verification and uncertainty assessments were carried out. The distribution of the tracer in the axial direction was strongly influenced by the injection site selection, which also had an impact on the AUC, Cmax, and Tmax values. In comparison with previous simulations, CM and ICV injections significantly increased the AUC to brain regions. In comparison to ICV, CM injection demonstrated higher AUC in the cerebral subarachnoid space (SAS). The ventricles were reached by both ICV and CM injections, although CM had a significantly lower AUC there. Tracer transport within the ventricular system was impacted by the CSF production rate. The effects of CSF production rate and treatments on solute transport inside the ventricular cisterns can be investigated in more detail in future research. It was discovered through a comparison of LP, CM, and ICV injections that steady-streaming flow patterns that are geometrically created at various vertebral levels have an impact on solute transport. For CM and ICV injections compared to LP, steady-streaming in the cervical spine is quicker than in the lumbar spine, leading to a quicker and more symmetrical tracer dispersion in both the caudal and cranial directions. Due to increased steady-streaming in the cervical spine, the initial peak tracer concentration close to the injection site fell more quickly with CM injections. LP injections, on the other hand, displayed a high concentration tracer zone for about 5 min following injection. Tracer dispersion was affected by parametric changes to injection techniques, though to a smaller amount than the injection location. Up to a 2X increase in intracranial AUC over the baseline, instance was seen with a 3X increase in bolus volume and the addition of a 5 mL flush. A 5X increase in bolus rate, however, had no positive impact on tracer exposure to the brain. The concentration differences at any given time and place did not surpass 10%, demonstrating the requirement for a high-precision in silico platform to identify such minute variations. Tracer transport to the brain was more strongly influenced by physiological factors, notably cardiac- and respiratory-induced changes in CSF stroke volume and frequency, than by changes in bolus rate, volume, or flush. Tracer distribution to the brain, basal cistern, and cerebellum was improved by increasing cardiac and respiratory CSF movement. Changes in injection methods showed less of an effect than differences in physiological variables, which may assist to explain experimental heterogeneity in non-human primate investigations. Our in vitro and virtual studies support the general conclusions and tracer transport kinetics found in earlier research. According to earlier research, bolus volume expansion and the addition of a flush can improve medication absorption into the cranium. In a similar vein, our research revealed that a 5 mL flush increased brain biodistribution by up to 3-5X above the initial value. The cervical spine's movement patterns and the timing of the tracer transfer along the spine matched up with earlier studies. Numerous MR imaging experiments using intrathecal drug injection in animals and people have been carried out. In terms of tracer arrival time and distribution in the cranial region, our results for tracer dispersion are consistent with earlier studies. The simulation results demonstrated minimal tracer migration to the ventricular cisterns after LP injection, which is consistent with earlier research. These results support the validity of our models as well as the possibility of more effective medication delivery to the brain via CM and ICV injections in comparison with LP. There were some limitations and potential research topics noted in this study. The CFD model did not use the dynamic mesh technique, and adding a glymphatic system and a non-uniform flow might make the model more thorough. Future research should address the lack of fluid–structure interaction (FSI) and physiological interactions related to connected systems. The impact of injection parameters such as injection angle and dorsal–ventral injection was not thoroughly investigated, and diffusion effects were also disregarded. Tracer absorption into the CNS tissue, the presence of arachnoid trabeculae, and in vivo validation of the results were not done. Other restrictions included the complexity of the CSF system and the absence of compliance differentiation in the tissue of the spinal cord and intracranial compartments. Investigation of the asymmetric tracer dispersion inside the skull, as well as variables affecting the geometry of the cortical subarachnoid space and spinal curvature, is necessary. This study looked into how different factors affected intrathecal solute transport inside the CSF system. The outcomes demonstrated that tracer distribution to the brain was most significantly influenced by injection location, particularly CM injection. Higher stroke volume and deep respiration dramatically increased tracer exposure in the cranial SAS, but bolus amount and pace had a less dramatic effect. The results demonstrate the potential impact of small changes to LP injections and underline the significance of selecting an injection location near the target area.

A new CFD model for convection-enhanced delivery (CED), a technique for delivering drugs to the brain, is also taken into consideration (Messaritaki et al., 2018). The consequences of diffusion non-Gaussianity are taken into account by the model's incorporation of diffusion probability as determined by diffusion MRI. The study shows that accounting for these effects enhances forecasts of drug concentration, especially in white matter regions. For further confirmation, experimental validation including gadolinium infusion on animals and phantoms is planned. For convection-enhanced drug delivery (CED) to the human brain, this study proposes a theoretical framework and a CFD model. The R-model, a unique model that accounts for non-Gaussian diffusion in brain tissue through diffusion MRI measurements, adds diffusion probability. The D-model, in comparison, is the prior model that makes use of the diffusion tensor. The R-model exhibits isotropic drug spreading in grey matter regions and anisotropic distribution following fibre directions in white matter regions, showing improved agreement with experimental observations. The need for diffusion non-Gaussianity in CED models is further supported by comparisons with the body of research since the R-model more closely matches reported drug distributions than the D-model in a variety of experimental settings. The final goal of CED treatment, diseased brain tissue, is highly sensitive to the effects of diffusion non-Gaussianity. Parkinson's illness, epilepsy, and Huntington's disease all have altered microstructure and diffusion properties, highlighting the significance of taking non-Gaussian diffusion into account when modelling CED for these patients. The study also discusses how extended infusion periods and injection into white matter structures may affect how drugs are distributed, highlighting the importance of taking diffusion non-Gaussianity into account in computer simulations. Even though the study lacks experimental validation, subsequent work will involve administering gadolinium to animals and brain-like phantoms to track drug distribution using MRI imaging. The authors are aware of the model's shortcomings, including the catheter's simplified modelling, the assumption of hard brain tissue, and the absence of clearance or metabolism effects. Overall, the R-model and its theoretical framework aid in the comprehension and improvement of CED for medication delivery to the human brain.

3.3 Ocular System

Fluids can be categorized as Newtonian or non-Newtonian substances that deform when subjected to shear stress. The numerical technique of computational fluid dynamics (CFD) is used to investigate fluid flow and other physical processes. For linearized potential equations to exist, the Navier–Stokes equations must first be simplified. Research, particularly computer modelling of blood arteries, can be revolutionized by multidisciplinary collaboration in domains like cardiology. Aqueous humour inflow and outflow are out of balance in the eye condition glaucoma, which causes elevated intraocular pressure and harm to the optic nerve. For a deeper comprehension of the characteristics of aqueous humour and its function in glaucoma pathogenesis, computational fluid dynamics is desired. The peculiarities of aqueous humour outflow in the human eye have been examined by researchers, who have identified several physical factors that contribute to these fluxes. These include, among others, buoyancy-driven flow, ciliary body-generated flow, gravity-induced flow when sleeping, flow brought on by phacodonesis, and REM sleep. It has been investigated how these fluxes affect elements such as pigment particle detachment, hyphemas, hypopyons, and iris contour using mathematical models and computer simulations. These investigations help us comprehend ocular fluid dynamics and how it affects eye health. In the context of glaucoma and laser iridotomy, CFD has been used to investigate the effects of shear stress on corneal endothelial cells (CECs). The results imply that aberrant aqueous flow and elevated shear stress, particularly in eyes with small anterior chambers, can cause CEC damage and loss. Understanding the aetiology of glaucoma and evaluating prospective treatments are both made easier by CFD. The vitreous, which resembles gel, occupies the area in the eye between the lens and the retina. Water, collagen fibres, hyaluronic acid, and proteins make up the majority of it. It helps keep the shape of the eye and has a gelatinous consistency. But as we age or develop certain eye problems, it might change, liquefying or detaching. Pharmaceutical medicines are frequently injected intravitreally to treat eye conditions. Drug distribution and dissolution in the eye have been investigated using computational fluid dynamics simulations. Studies have demonstrated that medication concentrations can be affected by injection settings and implant location but that the particle size of suspensions may not have a substantial impact on their performance. These models support enhancing therapeutic advantages and reducing tissue toxicity. The lens in the eye has been examined for heat exposure and damage using CFD. The amount of the lens's optical damage was shown to depend on how long it was exposed to heat, according to research. This study emphasizes how heat exposure might have negative effects on the lens epithelial cells. CFD can be used to study the eye's accommodation, which enables focusing on objects at various distances. Together, the ciliary body and lens modify the lens's shape and hence its focusing power. Age-related reductions in accommodation result in presbyopia. To research fluid dynamics during cataract surgery, including factors like irrigation/aspiration, microincisions, cataract grades, and microcoaxial method, CFD is being investigated. It helps to advance surgical methods and provides insights into fluidic properties.

The topic of ocular medication delivery is covered in the research, along with its main distribution methods, such as surface, intravitreal, subretinal, and subconjunctival administration (Dosmar et al., 2022). It examines each strategy's drawbacks and benefits while highlighting the promise of sustained-release medication delivery systems. To create efficient ocular medication delivery systems, the research also stresses the significance of taking anatomy, physical obstacles, important cells, and drug-biomaterial interactions into consideration. The cornea, conjunctiva, and tear film make up the eye's surface and are all essential for vision and immune protection. The conjunctiva lubricates the eye and includes immune cells, whereas the cornea serves as a barrier and refracts light. The tear film, which is made up of water, salts, and antibodies, preserves eye health and makes medication distribution easier. Drugs are delivered to the eye via a variety of techniques, each having advantages and limitations, such as eye drops, emulsions, suspensions, ointments, and contact lenses. The vitreous body, a gel-like substance that mostly consists of water, collagen, and hyaluronic acid, fills the intravitreal space of the eye. It has hyalocytes, fibroblasts, and macrophages as its cell types. The retina and optic nerve, among other eye tissues, are connected to the vitreous. Drugs are frequently delivered to the posterior portion of the eye through intravitreal injections and implants, with implants offering prolonged release and fewer systemic adverse effects. A target for medication administration in the treatment of retinal disorders is the subretinal space, which is situated between the retinal pigment epithelium (RPE) and the photoreceptive cells. Currently being investigated techniques include retinal prosthetics, subretinal injections, and transplants. While innovative delivery techniques including nanoparticles, liposomes, hydrogels, and altering the blood-retina barrier are being researched, gene therapy and cell therapy are potential treatments. Drug administration to both the anterior and posterior parts of the eye may be possible through the subconjunctival space, which is present between the conjunctiva and the sclera. To enhance drug dispersion and sustained release, drug delivery technologies such as liposomes, hydrogels, and polymeric controlled release systems are being investigated. Although there are difficulties with clearance, size, and degradation, these systems have advantages in terms of controlled release and targeted drug administration. Systems for delivering drugs to the eye that have a sustained release depend on drug carriers that can communicate with the ocular environment. Different substances have unique interactions and restrictions, including chitosan, alginate, PEG, PLGA, and NiPAAM. Alginate and chitosan can be used in neutral or acidic conditions, whereas PEG possesses solubility qualities but can also cause protein adsorption. Although PLGA enables multidrug delivery, it might obstruct antibody active sites. NiPAAM offers hydrogels temperature-dependent control over drug release. To create efficient sustained-release drug delivery systems for ocular applications, it is essential to comprehend these material interactions. Pharmaceutical anti-VEGF therapies include aflibercept (Eylea), ranibizumab (Lucentis), and bevacizumab (Avastin) and are used to treat a variety of ocular disorders. Eylea binds to several ligands and has a higher binding affinity for VEGF-A. Though their molecular sizes and tissue distributions are different, Avastin and Lucentis have comparable processes. These medications have demonstrated promising outcomes in the treatment of ocular diseases and VEGF inhibition. For particular ocular disorders, other medications are also used, such as Macugen and Visudyne. This paper gives a summary of the benefits and drawbacks of several ocular medication delivery methods. The difficulties of topical and repetitive dosing are being overcome through implants and sustained-release technology. As they go through clinical trials and receive regulatory approval, retinal prosthesis, and polymeric-based sustained-release devices are predicted to become more prominent on the market in the upcoming years.

The construction of in vitro and in silico models to examine medication delivery through the outer blood-retinal barrier (OBRB) for intraocular drug delivery is also considered for research (Davies et al., 2020). The models showed promise as predictive tools for researching drug transport in this tissue by correctly predicting the migration of dextran molecules and the release of ibuprofen from silicone oil. In the investigation, ibuprofen and fluorescein isothiocyanate conjugated dextran (FD-4) were utilized. With recirculating flows and fluid penetration through the ePTFE_M membrane, the computational models of drug transport through the outer blood-retinal barrier (OBRB) exhibited comparable velocity patterns for varied flow rates. Experimental data were used to validate the simulation results, which revealed an agreement with the in vitro results for dextran concentration profiles. The simulations proved the potential of merging computational and experimental approaches in understanding drug transport across the OBRB and correctly predicted the maximum concentration seen in the studies. Tracking the fluid–fluid interface was essential for analysing medication concentration dispersion. A meniscus formed and different flow fields within each phase were depicted by an adaptive mesh used to simulate the interaction between the silicone oil and aqueous phases. The simulation displayed consistency with experimental data and correctly anticipated the amount of ibuprofen in the receptor compartment. Computational models are useful for quickly, affordably, and accurately representing complex systems. They can also be applied to the analysis. Models can only approximate biological situations, but they nonetheless offer predictions for a range of factors. The research advances the possibility of a computational model of the eye as well as the development of a drug delivery system. Drug behaviour in complicated settings, such as the blood-retinal barrier (BRB), can be predicted using in silico models in conjunction with in vitro models. The field of drug discovery may gain from this strategy and the design of drug delivery systems may be improved. However, as no model can completely mimic the intricacy of the human eye, care should be taken when interpreting model results. Nevertheless, the congruence between experimental results and clinical behaviour can be enhanced by more sophisticated designs that imitate choroidal flow.

Researchers also studied bioengineering applications of CFD to deliver medications to the posterior region of the eye (Abootorabi, 2020). The first use optimizes drug concentration by adding a porous separator to the implant to target drug administration to the posterior eye. In the second application, wall shear stress and its effects on bone cells are studied by modelling pulsatile flow in a rectangular duct. The purpose of this study was to determine how implant structure affects medication delivery to the posterior eye. An implant was fastened to the sclera layer of a representational model of the human eye. The front surface of the eye was excluded due to its low drug permeability whereas the vitreous humour, a porous gel-like substance, is taken into account. In particular, drug diffusion in the posterior eye to cure age-related macular degeneration is the emphasis of this work's computational fluid dynamics (CFD) in the bioengineering discussion. In the study, medication implants are investigated as a possible safer drug delivery method than direct injections. Investigated are the results of inserting a porous separator to regulate drug distribution. A verified model was also used to study wall shear stress using the lattice Boltzmann method. However to improve drug delivery and computational accuracy at boundaries, more study is required.

Intravitreal injections of eye medications are visualized and predicted using microCT imaging and CFD modelling (Smith, 2011). Ex vivo pig eyes were used to trace the movement of a drug mimic, and the results offer exact details on time, flow patterns, and concentration. The created computer model improves our comprehension of drug delivery to the target site by precisely predicting drug transport. There are several ways to deliver drugs to the eye, focusing on either the anterior or posterior areas. While posterior delivery makes use of systemic administration, topical delivery, or direct injection, anterior delivery uses ocular drops for subconjunctival injection. Barriers like the cornea, aqueous humour turnover, and the blood-retina barrier present difficulties. This study focuses on intravitreal injections since they are frequently utilized to better understand the vitreous's drug transport characteristics and deliver medications to the eye's posterior portion. To evaluate X-ray compatibility and transport characteristics in ocular drug delivery, this work makes use of microCT imaging. The use of different contrast agents as medication mimics and tags includes iodinated contrast and gold nanoparticles. Analysis of signal strength, clearing time, and diffusion coefficients is possible with non-invasive imaging. To improve intravitreal injection methods, anomalies such as air bubbles and needle tip movement are investigated. The construction of a computational fluid dynamics model will help to better understand how drugs are distributed in the eye and will support the creation of innovative drug delivery techniques. About iodine and gold nanoparticles used as contrast agents in ocular drug administration, this work supports the linearity of signal intensity and concentration analyses. For every scan, calibration using diluted contrast agent vials is crucial. Due to their slower dilution rate and higher molecular weight, gold nanoparticles provide greater contrast at lower doses, making them a suitable medication surrogate. Additionally, they provide multimodal imaging through the ability to tag with different agents. However, taking into account other pharmacokinetic features, the presumption that these contrast agents respond similarly to therapeutic ocular medicines is a restriction. This study uses microCT imaging to show how a contrast agent bolus that mimics a medication can be tracked in real-time via the vitreous. The bolus evolution may be seen and evaluated quantitatively because of the high-resolution imaging. Iodine and gold nanoparticles' vitreous diffusion coefficients are calculated, revealing time-dependent effects and variations from predicted molecular weights. Limitations include the requirement for additional computational fluid dynamics methods to comprehend diffusion and advection transport in a finite volume, as well as the usage of contrast chemicals as drug imitators. It's crucial to keep air bubbles to a minimum while administering intravitreal injections to prevent transient vision loss and poor drug delivery. Inadequate de-airing, cavitation, and air entrainment during needle penetration are a few potential sources of air. Higher injection velocities and prefilled syringes can enhance drug delivery. Fluid velocity, collagen network instability, and needle insertion all have an impact on the injected bolus's form. Optimizing injection techniques and addressing issues like reflux can increase therapeutic efficacy. An innovative component of this study was keeping the needle in the eye while imaging. Based on substantial experimental data, a computational fluid dynamics (CFD) model was created to mimic intravitreal medication delivery. The model, which provided spatial distribution data and concentration profiles within the vitreous, had good agreement with ex vivo data. Air bubbles and different bolus forms were studied using the model, which revealed their insignificant effects on drug concentration gradients. The model's ease of use and adaptability make it appropriate for conversion to human models, bridging the divide between testing on animals and people in drug delivery investigations. To look at the flow and distribution of medications in the vitreous after intravitreal injections, this study combines microCT imaging and computational modelling. Imaging techniques were used to determine the concentration profiles of gold nanoparticles, which were discovered to be an appropriate medication mimic. Ex vivo models and CFD models were created to shed light on the mechanisms underlying drug transport and unusual characteristics. The research paves the way for more in-depth 3D CFD simulations of ocular medication flow.

Clinicians have started prescribing therapeutic lenses that contain the required medication to improve drug delivery to the eye (Silva et al., 2012). If therapeutic glasses or topical eye drops are more helpful in this regard, it will be shown mathematically in this study. The mathematical model utilized in this work takes into account the following:

(i) The diffusion processes take place in the anterior chamber, cornea, and therapeutic lens.
(ii) The drug's metabolic breakdown in the cornea and anterior chamber.
(iii) The convection processes brought on by the aqueous humour's movement within the anterior chamber.

The study seeks to compare the efficacy of topical eye drops versus therapeutic lenses and provide a thorough understanding of medication delivery efficiency by including these parameters in the model.

All of the models in this study account for trabecular mesh blockage, which causes a higher intraocular pressure (IOP) of 3600 Pa. It is crucial to remember that the normal IOP in a healthy eye is usually around 1950 Pa. The study's objective is to evaluate the effect of raised pressure on the effectiveness of drug delivery to the eye using either topical eye drops or therapeutic lenses while taking into account this elevated IOP. The study concludes that therapeutic lenses are superior to topical treatment for glaucoma. Higher medication concentrations are achieved and maintained in the eye for longer when therapeutic lenses are worn. The researchers are currently working on putting into practice a more accurate model that depicts the trabecular mesh as a porous media. In addition, they are running simulations to assess the therapeutic impacts of different medications on intraocular pressure (IOP) in the management of glaucoma.

3.4 *Transdermal System*

Transdermal drug administration is an appealing option to oral drug delivery and has the potential to offer a substitute to hypodermic injection as well. People have been applying chemicals to their skin for countless years for therapeutic purposes, and in the modern period, a wide range of topical applications have been produced for use in local indications. Hydrophilic compounds, macromolecules, and vaccinations can all be administered transdermally, thanks to a variety of transdermal delivery techniques. There are several different transdermal medication delivery methods, including passive, iontophoresis, skin abrasion, thermal ablation, metered-dose transdermal spray, chemical enhancer, and microneedles (Prausnitz et al., 2008).

A non-invasive drug delivery method known as a needle-free injection system is rapidly expanding its uses beyond the administration of vaccinations and insulin. A needle-free injector produces a fast speed microjet stream of a drug, typically at a speed greater than 100 m/s that enters the skin and tissue barrier to deliver the drug at a specific depth. For the objective of understanding the fundamental physics underlying the generation and dispersion of microjets during injection procedure, experimental research and following computational fluid dynamics (CFD) analysis for an air-powered needle-free injection system for dermatological purposes were conducted (Mohizin et al., 2018). This study's objectives were to comprehend the physics underlying microjet generation and propulsion in an air-powered needle-free injectors and to pinpoint the crucial variables, such as filling ratio, driving pressure, nozzle diameter, and fluid type. The skin surface's stagnation pressure was employed as a quantifier of the injection's success. The experimental data could be reasonably predicted by the computational data, and this information may be utilized to construct and improve air-powered needle-free injectors.

Researchers performed research to demonstrate the effects of lidocaine, an analgesic drug, through a microneedles array and to evaluate how the fluid would be transported across the skin's transcellular routes using CFD simulation techniques. The investigation was complemented by a CFD-based computational simulation which closely approximates the action of lidocaine in specified scenarios. Two microneedle array-based designs are examined in particular, representing an internal flow that is lateral as well as an internal vertical flow. The flow distribution in the microneedles

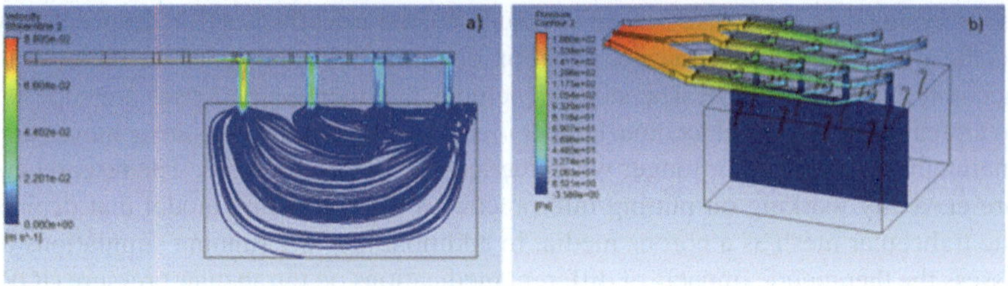

Fig. 4 CFD simulation for administering analgesics: **a** Velocity and **b** pressure profiles of drugs flowing from the microneedle array to the skin (Henriquez et al., 2023)

array and the medium that is porous are considered independently in order to evaluate the uniformity in the proper distribution of a pain reliever to each microneedle, and then to simulate the functioning of the joint situation in order to evaluate the impact of the change within the microneedles array (Fig. 4). They have concluded that to avoid turbulence at medication entrance, the management of factors like velocity and pressure required for the deployment of microneedles array-based devices should be studied (Henriquez et al., 2023).

References

Abootorabi, S. (2020). *Computational fluid dynamics for modelling and simulation of intraocular drug delivery and wall shear stress in pulsatile flow* (Doctoral dissertation).

Davies, A. E., Williams, R. L., Lugano, G., Pop, S. R., & Kearns, V. R. (2020). In vitro and computational modelling of drug delivery across the outer blood–retinal barrier. *Interface Focus, 10*(2), 20190132.

Dosmar, E., Walsh, J., Doyel, M., Bussett, K., Oladipupo, A., Amer, S., & Goebel, K. (2022). Targeting ocular drug delivery: An examination of local anatomy and current approaches. *Bioengineering, 9*(1), 41.

Henriquez, F., Celentano, D., Vega, M., Pincheira, G., & Morales-Ferreiro, J. O. (2023). Modelling of microneedle arrays in transdermal drug delivery applications. *Pharmaceutics, 15*(2), 358.

Heyder, J. (2004). Deposition of inhaled particles in the human respiratory tract and consequences for regional targeting in respiratory drug delivery. *Proceedings of the American Thoracic Society, 1*(4), 315–320.

Jooybar, E., Abdekhodaie, M. J., Farhadi, F., & Cheng, Y. L. (2014). Computational modelling of drug distribution in the posterior segment of the eye: Effects of device variables and positions. *Mathematical Biosciences, 255*, 11–20.

Khani, M., Burla, G. K. R., Sass, L. R., Arters, O. N., Xing, T., Wu, H., & Martin, B. A. (2022). Human in silico trials for parametric computational fluid dynamics investigation of cerebrospinal fluid drug delivery: Impact of injection location, injection protocol, and physiology. *Fluids and Barriers of the CNS, 19*(1), 8.

Kleven, M., Melaaen, M. C., Reimers, M., Røtnes, J. S., Aurdal, L., & Djupesland, P. G. (2005). Using computational fluid dynamics (CFD) to improve the bi-directional nasal drug delivery concept. *Food and Bioproducts Processing, 83*(2), 107–117.

Longest, P. W., Bass, K., Dutta, R., Rani, V., Thomas, M. L., El-Achwah, A., & Hindle, M. (2019). Use of computational fluid dynamics deposition modelling in respiratory drug delivery. *Expert Opinion on Drug Delivery, 16*(1), 7–26.

Messaritaki, E., Rudrapatna, S. U., Parker, G. D., Gray, W. P., & Jones, D. K. (2018). Improving the predictions of computational models of convection-enhanced drug delivery by accounting for diffusion non-Gaussianity. *Frontiers in Neurology, 9*, 1092.

Mohizin, A., Roy, K. R., Lee, D., Lee, S. K., & Kim, J. K. (2018). Computational fluid dynamics of impinging microjet for a needle-free skin scar treatment system. *Computers in Biology and Medicine, 101*, 61–69.

Morita, K., Takeishi, N., Wada, S., & Hatakeyama, T. (2022). Computational fluid dynamics assessment of congenital tracheal stenosis. *Pediatric Surgery International, 38*(12), 1769–1776.

Prausnitz, M. R., & Langer, R. (2008). Transdermal drug delivery. *Nature Biotechnology, 26*(11), 1261–1268.

Silva, P. M., Ferreira, J. A., & de Oliveira, P. (2012). Tool for studying drug delivery to the eye in case of glaucoma. *Transfer, 4*(5), 1.

Smith, C. A. (2011). *Development of imaging paradigms for drug distribution and fate in the eye.*

Srivastav, V. K., Kumar, A., Shukla, S. K., Paul, A. R., Bhatt, A. D., & Jain, A. (2014). Airflow and aerosol-drug delivery in a CT scan based human respiratory tract with tumor using CFD. *Journal of Applied Fluid Mechanics, 7*(2), 345–356.

Taherian, S., Rahai, H., Gomez, B. Z., Waddington, T., & Bonifacio, J. R. (2015, November). Tracheal stenosis: A CFD approach for evaluation of drug delivery. In *ASME international mechanical engineering congress and exposition* (Vol. 57380, p. V003T03A096). American Society of Mechanical Engineers.

Tiwari, A., Jain, A., Paul, A. R., & Saha, S. C. (2021). Computational evaluation of drug delivery in human respiratory tract under realistic inhalation. *Physics of Fluids, 33*(8).

Tiwari, A., Paul, A. R., & Jain, A. (2022, December). Challenges in drug delivery through intranasal route: A CFD perspective. In *conference BioSangam 2022: Emerging trends in biotechnology (BIOSANGAM 2022)* (pp. 35–46). Atlantis Press.

The manufacturer's authorised representative in the EU is Springer
Nature Customer Service Centre GmbH, Europaplatz 3, 69115 Heidelberg,
Germany. If you have any concerns regarding our products, please
contact ProductSafety@springernature.com

Printed and bound by CPI Group (UK) Ltd, Croydon, CR0 4YY
30/09/2026
02225324-0002